To all my children and grandchildren

Your future is my inspiration!

sustainable GARDENING MADE EASY

FROM DESIGN TO HARVEST: HOW TO GROW ORGANIC, SUSTAINABLE FOOD IN COLD CLIMATES

Permaculture explained

How nature works when left alone

From plans to plants to harvest: what to do

EVA PETTERSSON

Translation of the swedish book "Lätt att odla naturligt": Madelene Dahlberg KOMUDESIGNS www.komudesigns.com
Cover & Layout: Madelene Dahlberg KOMUDESIGNS www.komudesigns.com
Illustration: Eva Pettersson
Photo: Pixabay, Pexels och Shutterstock

Järbo
OM Förlag
2021
ISBN: 978-91-984965-1-2

Introduction

Today, most people live in cities and communities with less and less contact with nature. Just 150 years ago — reality looked very different. Back then, most people lived in the middle of nature and knew how it behaved. During that period in Sweden, seven out of ten people worked in agriculture. Today, only about two per cent do. We have in a very short time gone from an agricultural society to an industrial society to today's service-based society. Fortunately, the knowledge of those days is still with us, in our oldest generation. This knowledge is something that we need to help preserve before it disappears completely.

It is with a tremendous and passionate commitment to nature, agriculture, and the food on our tables that I have written this book.

I will, therefore, give a brief overview of what today's food production looks like, its pitfalls, and what is needed for us to move on to sustainable food production.

I do this because of my heartfelt desire for more people to understand how unnaturally things are grown today, and how important - and easy - it is to grow naturally.

Food production today

Today's industrial farming, with fewer and fewer farms with ever-increasing monocultures, causes enormous problems - such as deforestation, soil erosion, and pollution. Fresh ground is continuously being broken to grow the amounts of corn, soy, grains, and oilseeds that the industry is demanding for food, animal feed, and fossil fuel. Here are some of the big problems in modern agriculture:

→ *Very high energy consumption. It takes 10 calories of energy to produce 1 calorie food.*

→ *A third of our soils are already depleted. If today's level of depletion continues, there may be as little as 60 years of nutrient-rich soil left.*

→ *A harsh hunt of new land, known as land grabbing, is taking place in various developing countries - especially in Africa.*

→ *Unsustainable irrigation that causes the planet's freshwater to deplete, or be polluted quickly.*

→ *The extensive use of chemicals, from fuel and fertilizers to various forms of pesticides - most of which comes from the oil industry.*

→ *Vegetables contain fewer nutrients as they get less nutrition themselves in the form of artificial fertilizers.*

Large-scale animal production

Production of various animal products such as beef, pork, poultry, and fish is also heavily industrialized. Here are the major problems:

→ *Very high energy consumption.*

→ *The animals live in very unnatural conditions.*

→ *The animals have abnormal diets, such as soy and corn.*

→ *Feed and even to some extent water must be transported to the farms as the animals do not have access to any natural food.*

→ *Vast amounts of manure become a waste issue, rather than valuable fertilizer.*

→ *This meat contains fewer nutrients, including lower levels of omega 3.*

Large-scale organic crops

Just because a farm has gone organic does not always mean that it is better or more sustainable than the conventional farm was.

Organic systems can be as unsustainable as the conventional ones if these are made up of large monocultures. In many greenhouses, crops are not grown in real soil - but in various substrates where all nutrition comes from the irrigation with different nutritional solutions

The only good thing about these large-scale organic crops is that the toxic chemicals are gone.

Modern biochemistry

There is a silent change happening with our nature and our food that is spearheaded by the biochemistry industry, where intensive research is underway to produce new effective products of various kinds. All to increase returns; especially financially.

This is done using everything from newer forms of gene modification (GMO) to the latest big thing, known as "synthetic biology".

Genetically modified plants and animals

The most common form of gene modification today is the so-called molecular scissor CRISPR. Today, the researchers cut and paste DNA in everything from bacteria, yeast, and green plants to zebrafish, fruit flies, mice, rats, and human cells.

Synthetic biology

The newest thing is something called synthetic biology. It is a form of genetic modification that is based on creating new artificial lifeforms, in which nature no longer decides what and in what form something should exist. Instead, the industry's own assessments determine this.

Today's food supply

Everything we've decided that we need is produced far away from us and transported to the cities - just as the shelves need restocking.

A severe effect of today's food industry is that most people have lost all knowledge and understanding of how things are produced. There are frightening many of us who do not know how our food is produced - much less what it contains.

When both vegetables and animal products are produced at the lowest possible cost with maximum profit, no consideration is given to what the food ultimately contains. The problem is that plants and animals are what they eat! On top of this, more and more food products are eked out with everything from water to various chemical additives.

Domestic food supply is said to sit at just under 50%. This may not sound that alarming, but you need to remember that everything produced in Sweden is entirely dependent on imports of so-called input goods, such as fuel and fertilizers. The fuel sector has some domestic stocks, but in a crisis, these reserves won't last more than a few days.

In practice, we have become entirely dependent on transports from all over the world working around the clock. This is an unsustainable system that risks collapse at the slightest disruption of the various transport routes.

Towards a more natural society

Today, most people know that much in society needs to change in order to have a sustainable future. Unfortunately, too little is done, and many people feel that it's too negative, too difficult, and almost impossible for them to make a difference. But that is where reality differs!

Just one person can do a lot, and the best part is that once you start doing something - it starts to feel less negative. As an inspiration along the way, I would like to briefly tell you about the few indigenous people who still live with nature.

→ *They see man as part of the earth.*

→ *They make no difference between animals and plants, but only see animals as a higher form.*

→ *Their cultivations are similar to forest gardens.*

→ *Their gardens are comparatively small, but provide the people with everything they need; food, some medicines, wood, fibres, and even a small surplus to exchange, give, or sell.*

→ *The size of a community is about one hectare, and fascinatingly enough, whether the village is in Africa, Asia, or South America - the size remains the same.*

Organic and small-scale farming

Some time ago, the UN issued a report that points to the problem of today's industrial agriculture.

In the report, they say that the world can cope with the challenges in terms of food supply, fewer chemicals, and less fossil fuel that we face. They mean that by converting to local, ecological, small-scale, and sustainable agriculture, we will be able to support an increased population. At the same time, nature and the environment will thrive.

In organic farming, no toxic chemicals are replaced by natural nutrients such as green manure and animal manure. The result is healthy and fertile soils that yield crops of non-toxic and nutrient-rich food.

The best form of organic farming is one where you grow with a crop rotation schedule and keep different kinds of animals. Such a system does not deplete the soil, but rather strengthens it.

Animals are not only needed for food, but they are also required first and foremost for the work they do in the fields. When they graze, they fertilize the soil with their manure and keep the soil surface porous with their cloven hoofs. This is important, because when the soil is porous - the rainwater is absorbed by the soil and does not flow off the ground as it does in those vast monocultures. This, in turn, reduces soil erosion.

In an organic system, the animals are also allowed to live more naturally and eat the food they were intended to eat. Which, in turn, means that they stay healthier and that less medication is needed. Good for the animals, but also for those of us who choose to eat their meat. Here too, the result is a non-toxic and nutrient-rich food.

How can you and I help?

The hardest part can be knowing where to start, but since money is what controls today's food production, we can use this to our advantage! As consumers, we have a more significant influence than we think. When profit is the deciding factor, the products we buy the most of are also the ones that are produced. By changing our buying habits, we can help break the industrial circle. Ultimately, we deceive ourselves with the beautiful exterior of nutrient-poor products. Organic apples might not always be as pretty, but they taste much better!

An excellent way to start is to find out what the food you are buying contains, and then choose the option with the purest ingredients and the least amount of additives. This is best done by consistently choosing food that is organic, local, and produced on a small-scale. This means that we cannot always get hold of all those things we are used to buying, but remember the old saying "Good things come to those who wait"? As it happens, it's not wrong to wait for something tasty - when the season for it finally comes, it'll taste all the sweeter!

As a consumer, it is time to put your foot down and choose the food that makes both us, the animals, and nature happiest!

Start growing your own

Starting to grow your own food is not only exciting and fun, but it also yields food that you know is healthy. In addition, it provides you with the satisfaction of contributing something positive to the change that needs to happen.

If you don't know where to start, I would highly recommend permaculture. Around the world, the number of people who use permaculture with great success is steadily growing.

Initially, permaculture may feel a little alien, and some of the ways of thinking may be perceived as being contrary to what we are used to. Still, once you understand the foundation - a world of crystal-clear simplicity opens up. All you need to know already exists in nature.

We need to slow down a little and start to follow nature's own rules and rhythms. After all, all things don't - and shouldn't - happen right away. Taking the time to observe, to think about what is available, and what is happening in your little piece of nature, provides not only knowledge - but also more opportunities to simply enjoy what you already have.

Then the book begins

The first part of this book is a review of how different biological systems work when nature is allowed to set the rules of the game.

The second part describes how permaculture works as a planning and cultivation tool.

The third part is the practical part, based on the first two parts of the book. Here you'll find lots of practical examples and illustrations, all to make it easier for you to get started with your crops.

There is also an appendix that contains many useful charts to help you along the way.

HOW *nature* WORKS WHEN LEFT ALONE

Photosynthesis – the key to life

Thanks to the green plants' ability to transform solar radiation energy into carbohydrates and oxygen, us humans, animals, and fungi can also take part in benefiting from solar energy. Without plants, we would not exist.

A vital process

In all green plants, two genuinely vital processes are working together simultaneously. One is the photosynthesis that captures energy from sunlight, along with carbon dioxide and water. Transforming the solar energy into plant energy in the form of glucose and into oxygen that the plant then releases to the air. The other is cellular respiration, which maintains the life of the plant when it is dark outside. The way cellular respiration works is that the plant absorbs oxygen from the air. This is, together with the glucose, converted into plant energy, water, and carbon dioxide, which is released into the environment.

Yellowing of the leaves

That the tree's leaves yellow and fall is due to insufficient light and water for the tree to be able to perform its photosynthesis. The tree withdraws its' building blocks of chlorophyll into its' trunk and roots to wait for the return of both light and water in spring.

Reverse photosynthesis

When we light our fireplaces, most people probably do not think about how the burning of the firewood creates the opposite of photosynthesis. Within the wood, solar energy that consists of carbon and water is stored as carbohydrates. When the wood burns, those carbohydrates react with the oxygen in the air to create carbon dioxide. The water within the wood becomes vapour, leaving behind ash. That ash contains the minerals that the tree absorbed from the ground during its' lifetime. The heat that the fire emits is the solar energy that the tree absorbed through photosynthesis, and some of that solar energy becomes the light that emanates from the fire.

Coal and oil; a controversial topic

The same principles apply to the burning of coal and oil. One could say that they filled with old carbon dioxide stored for millions of years, deep within the earth. Now, when all this energy is extracted and burned, this carbon dioxide is released – and there are not enough plants to "eat" the excess of it. When the earth's vegetation – especially the trees – decrease like today, and the human population continues to increase, the equation does not converge. The pattern of burning more fossil fuel than the plants can take care of must be changed, and everyone can help in some way.

In this, us gardeners, farmers, and growers – big or small – can contribute further to reduce the problem, provided that we do it sustainably, with nature as a guide.

A great example is that when we grow more of our food, most of that food's negative climate impact disappears.

Freshwater is no longer a given thing

Warnings about water shortages can be heard increasingly often. Yet the amount of water on our planet is the same as it has always been. The water shortage we hear about is regarding clean water, not water as a whole. We have a rapidly growing problem in that there is less and less fresh water available. But if everyone helps out, the turn of events can be halted.

Growing water scarcity

As it stands today, it's only when the water disappears or poses a problem that we turn our eyes towards the issue. If nothing happens, we take for granted that water will always flow from our taps upon turning them. However – this is a reasonably new mentality for us humans, and in many places, that mentality doesn't exist at all. In regards to Sweden, it's enough to go back half a century to see how much more careful we were with our water. During that time, showering daily was just not done. The bathtub was partially filled once a week, and the kids took their turn to wash up. And no one was dirtier then than we are now. Somewhere along the road up to today, we have lost respect for clean water. We need to learn to once again become adept at managing and respecting our water as the survival element it is.

I hope that this chapter, as well as the rest of the book, will instil a better understanding of how the water on earth is connected, and how it moves in an eternal cycle. And above all, how the water affects us, and how we affect the water.

How much water is out there?

Our oceans are humongous, and also where we have most of our water. A whopping 97 per cent of the earth's water is saltwater.

Those remaining THREE per cents are divided in:

→ *75 per cent frozen water in ice and snow*

→ *13.5 to per cent deeply buried groundwater, also known as fossil water, which lies at a depth of more than 800 metres.*

These two types are challenging to replace or replenish.

→ *11 per cent shallow groundwater, less than 800 meters down*

→ *0.3 per cent in lakes and ponds*

→ *0.6 per cent in the soils*

→ *0.03 per cent in rivers*

→ *0.035 per cent in the atmosphere*

These five forms, scarcely 12 per cent of the total 3 per cent, are the only water sources that we can influence.

The water's turnover period is the time it takes for all the water within a system to be replaced. Depending on which water system we're talking about, the turnaround period varies substantially. In the sea, the turnover period is an unbelievable 37,000 years. At the same time, it only takes a measly 14 days for the water in our bodies to be replaced.

Earth's freshwater

All living things need water. In agriculture, a lot of farms use irrigation systems. For organic farming that utilizes water from sources passing by their farms in a natural way, problems rarely occur. The water does its' job, sometimes several times over on the same farm. After which it flows on to the next area, where it can be used again since chemicals haven't destroyed it. Another positive example of irrigation is when you take wastewater – that hasn't been contaminated – from one business to another.

However, as it is right now, considerate water usage is implemented far too little, which means that the earth's freshwater supply is about to run dry.

Instead, what we have is more and more water that is too toxic to use, and often impossible to treat. We have ended up in a situation where some of our water has become hazardous waste.

Unfortunately, there is an infinite number of examples of how freshwater is abused and dirtied by anything and everything from our own toilets to various industrial processes as well as examples of water used solely for pleasure in irrigation of lawns and other, nonproductive areas.

The water's pH

Water pH plays a significant role in the processes of everything living. Naturally, – the soils pH levels are dependent on the water pH. A neutral pH is 7. The lower the number, the more acidic the environment, and the higher the number, the more basic.

Generally, rain is at a pH between 6.2–6.5, which would make it acidic. This means that the water in nature is also more acidic. One way to see this is when

water in nature is frothing. This is because acidic water froths more easily than alkaline (basic) water.

That our water has become more acidic is due to the different emissions into our atmosphere, and the more acidic the water, the higher the damage it can cause. Sometimes the rain is exceptionally acidic, and we call this acid rain. During this type of rainfall, the rain causes our seas, lakes, and lands to acidify. This has, in turn, caused fish to die, – has even caused whole bodies of water to die – and in nature; causes the leaves to yellow and dry out, just like in autumn. The plants' ability to absorb nutrients from the ground is also affected, as the acid rain causes the roots of the plants to die.

Artificial rain

We can create artificial rain, and this is already in play in many places around the world, through spreading small particles of silver iodide in a suitable area of rain clouds. The technique is far from safe and is up for debate more often than not. But the creation of rain is not really a new practice. The fact is that the rain dances of the various indigenous people may very well create rain if the right conditions are met. When they dance, a lot of dust swirls up, and if this dust is brought upwards with the wind, rain is a strong possibility.

Changing for the better

Change is needed in everything from a global perspective to private use.

Globally, more work needs to be done to reduce and clean up the emissions that make rain acidic. Then, the unsustainable deforestation has to cease, which I will discuss more in-depth in the next chapter; "Without trees, we would have deserts".

It is particularly essential to preserve the forests in coastal areas since trees desalinate seawater. Even in mountainous regions, the forest needs to be left alone as these trees collect rainwater which then benefits the lifeforms living down below.

Despite the somewhat hopeless picture, there is, in fact, a lot we can do and as the saying goes; "many a little makes a mickle". We can be better at reducing our water consumption, and we can learn to save and to reuse water in different ways.

Farther into the book, I will provide several examples of this, ranging from gathering water in rain barrels to acquiring cisterns, building smarter garden beds, and building ponds.

Without trees, we would have deserts

Trees are absolutely magical. They are singularly one of the essential parts of our ecosystem, as trees make up about 80 per cent of the workforce when it comes to creating soil and atmosphere. Without them, our climate would be completely different. The harsh reality is that without trees, there would be nothing but desert.

The water cycle of the tree

Thanks to a large leaf area, a fully grown tree collect massive amounts of water. A mature tree can have an incredible 16 hectares of working leaf area – if you count both the outer surface and all the vessels within the leaves.

Trees are also shaped in such a way that when it rains, the water runs along the outer branches towards the trunk, and down along it to the ground. Water will not drip from the crown until all the parts above ground have been saturated. When the droplets finally hit the ground,

it will be carrying nutrients from the surface of the tree along with insects and dust. This nutrient-rich water feeds the topsoil and the finer, shallower roots, while the water from the trunk feeds the deeper layers of soil and the more massive roots.

These thick, deeply buried roots pull different minerals from the ground – minerals which then gets transported up through the tree to the leaves. Once there, these nutrients are washed back down to the soil when it next rains. And so, the cycle goes. You could say that the tree feeds both itself and the ground surrounding it. There-

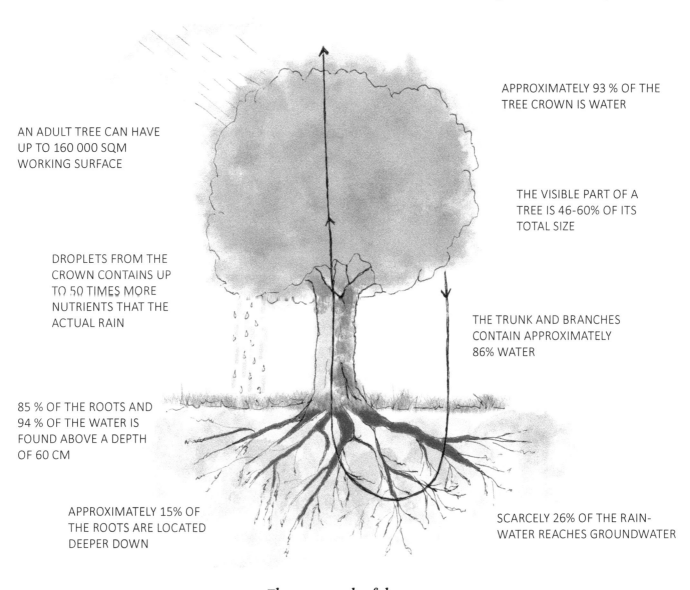

APPROXIMATELY 93 % OF THE TREE CROWN IS WATER

AN ADULT TREE CAN HAVE UP TO 160 000 SQM WORKING SURFACE

THE VISIBLE PART OF A TREE IS 46-60% OF ITS TOTAL SIZE

DROPLETS FROM THE CROWN CONTAINS UP TO 50 TIMES MORE NUTRIENTS THAT THE ACTUAL RAIN

THE TRUNK AND BRANCHES CONTAIN APPROXIMATELY 86% WATER

85 % OF THE ROOTS AND 94 % OF THE WATER IS FOUND ABOVE A DEPTH OF 60 CM

APPROXIMATELY 15% OF THE ROOTS ARE LOCATED DEEPER DOWN

SCARCELY 26% OF THE RAINWATER REACHES GROUNDWATER

The water cycle of the tree

fore, collecting rainwater under a tree gives you water that is significantly more nutrient-rich than just rain. Up to 50 times more nutritious than regular rain, to be precise!

The upper layer of soil and plant material underneath the tree – the so-called litter layer, or the O horizon – helps to retain the water on its way down through the ground. The first 60 cm of the soil below the tree includes 15 per cent of the tree's biomass (the roots) and 96 per cent of its water. This delay gives enough time for the roots to be able to take full advantage of the water before it continues down into the ground. Once the ground is fully saturated, and the water has given life to everything in its path, the water continues down to the groundwater.

Different types of soils have different capabilities for retaining water. Roughly speaking, a soil with no trees can hold 2.5–7 cm of rain per every 30 cm of depth, while the ground underneath a tree can hold up to 10–30 cm in the same volume of soil. Soil which is rich in organic matter, also known as humus, can hold up to one-third of its weight in moisture. One could say that figuratively, there exists a whole lake of recycled water within the forest's layers of soil.

Interestingly enough, trees supply more of the water present in streams compared to what the rain does. This means that the water in a brook is cleaner than rainwater – assuming that the environment around the brook isn't polluted.

Trees affect the climate

A tree recycles roughly 74 per cent of rainwater. This occurs partly by water evaporating from the surface of the leaves, and partly by water from the ground; water that travels up through the tree, where it then evaporates. The rest – almost 26 per cent – of the water flows down into the ground to become groundwater.

From this evaporated water, or rather the water vapour, clouds are formed. However, before a cloud can produce rain or snow, the moisture must adhere to the surface of tiny, tiny little particles. Most of which are minerals, although airborne microbes also help out. In this, the trees play another vital role, as water vapour from trees contains large amounts of those same little water-creating particles.

Clouds formed over forests are a mixture of water vapour from the sea and water vapour from the trees, depending on how far it is to the sea, or to a lake. Along the coastline, the rain contains 100 per cent seawater, while further inland, as much as 50 per cent of the water in the rain comes from trees.

Unlike water vapour from seas and oceans, water vapour from forests contains significantly more organic particles and plant nutrients.

The farther you get from the sea, the trees ability to condense moisture plays an increasingly important role. A long way inland, where you are far away from any seawater, there would be no rain at all without trees.

Another effect on our climate is that trees affect temperature; because trees – and other plants – are in themselves several degrees warmer than the surrounding air. During the day, this leads to evaporation that in turn cools the tree down, which then makes its' surroundings colder. The opposite happens at night, when the water condensates against the warmth of the tree, making its' surroundings warmer than the air nearby.

This, in turn, means that when hot, dry air enters a forest, it is cooled and moisturized. Vice versa, when cold, humid air enters a forest, it is heated up and dehydrated for it to evaporate on its way up through the canopies. Trees are therefore excellent at regulating both temperature, and moisture.

Trees work together with the soil and mycorrhiza

During their lifetime, trees produce their weight in organic waste several times over, which means that the trees produce most of their soil themselves. This aids the nutrient interaction – also known as mycorrhiza – between the plants' fibrous root system and certain sponges (Mycelium). These fibrous roots can, in the right conditions, extend an unbelievable 1000 metres away and 1500 metres down into the ground, which means that a tree can have a massive area to collect nutrients from. Concurrently, the fungi receive the starch they need from the trees.

Trees as air purifiers

Trees are our planets air purifiers that work just by living and thriving. They take care of the carbon dioxide and in return, gives us vital oxygen. In addition to carbon dioxide, trees also absorb other harmful gases such as carbon monoxide, chlorofluorocarbons, and nitrous oxide. Therefore, having large amounts of trees in our cities would be at least one way of reducing smog and other pollution problems.

With that in mind, it's easier to understand why the massive deforestation that is taking place across the planet contributes to the acceleration of climate change. There are simply not enough trees to convert the surplus of carbon dioxide to oxygen. So, in parallel with a massive reduction of emissions, we should be planting substantial amounts of trees, – urgently – which in the long run may help to curb climate change.

The effect of wind on the trees

Trees in the edge of the forest are both shorter and stockier than those farther in. This is the trees' way of "defending" themselves against the wind pressure that occurs when the wind hits the forest edge. Most of the wind is pushed up over the treetops to travel over the forest and falls in downward spirals once it reaches the other end of it; spirals that here in the northern hemisphere rotates to the right. If there is moisture and rain creating particles in the wind, rain is generated.

The rest of the wind passes through into the forest, where the friction against the trees forces the wind to rapidly decrease. Due to this, all the nutrition that was carried by the wind drops from the air only 100 metres into the forest. The friction also heats the forest, thus lowering the risk of frost. If you travel 1 km into the forest, you will notice that there is no wind! Here, the wind - figuratively speaking – has been transformed into wood.

When you are out and about in nature, it is interesting to see how the wind affects the trees, by observing the treetops. How much a tree crown leans and in what direction is determined by how powerful the winds are and what the dominant wind direction is. The effect of the wind is also visible on the growth rings of a tree since these are thinner on the windier side.

Using trees as climate protection

Everything I have described here makes it easier to understand how incredibly essential trees are to our climate. And that our trees, in many different ways, can help counteract the ongoing climate change.

Trees are an excellent tool for reducing CO_2 emissions; as a fully grown, normal-sized tree can "take care of" about 6 kilos of CO_2 per year. This may not sound like much, but in a forest, there are very many trees.

Unfortunately, it's not enough to merely plant more trees to stop climate change. The unsustainable deforestation must also be stopped. The ways we usually use nature must become more sustainable, both in the short term and in the long run.

Trees prevent flooding and soil erosion

The trees' ability to capture and store large quantities of water, both in their roots and in the tree itself, make trees excellent protectors against floods. Trees also protect against soil erosion, in that their extensive root system keeps the soil in place.

The more bare and impervious surfaces there are, the more difficult it is for the environment to deal with heavy winds and heavy rainfall. In places with no trees, the rain and wind are free to wreak havoc, which threatens to flood cities and easily washes away the soils of our fields.

A lot could be won if towns and cities replaced as much as possible of their paved surfaces with trees and other greenery. Likewise, the farmland's problem with soil being washed away could be significantly smaller if green zones of both trees and other plants were interspersed between fields.

Versatile soil for a happy environment

Without healthy soils, we cannot survive. In the soil, there are more than 50 million species of bacteria and more than 50 million species of fungi. So far, we have only named one per cent of these, and our knowledge of this one per cent is limited.

How nutritious soil forms

In nature, processes that create new fertile soil are always underway in environments such as:

→ *Shallow marine environments: marshes, mangrove swamps, riverbeds, and coastal waters.*

→ *Shallow lakes and ponds: Where large amounts of organic matter from different places gather. Imagine a gutter that hasn't been cleaned – plants quickly take hold and soil forms. Even if submerged or subaqueous soils aren't the same as soils on dry land, they still work the same when the water disappears.*

→ *The forest is continuously acting as a single, massive compost.*

→ *Grasslands: These become more fertile when animals graze an area. The soil is oxygenated thanks to the animals walking on it, and it is fertilized by the manure they leave behind. The winds bring heaps of nutrients that end up in all the little crevices of the ground, like seeds, for example; seeds that then grow to feed all the things living there.*

→ *Compost-based horticulture: These create lots of good, nutritious soil. With this technique, everyone can do their part in regenerating nutritious soil. As an added bonus, the food becomes more nutritious as well.*

Prerequisites for good soil

In order for the soil to be healthy and thriving, the following components must be in balance.

Water

The water contained in the ground – either from rain or irrigation – manages the transportation of water-soluble nutrients to the plants for their roots to access.

Even if you don't want your lands waterlogged, it is almost always preferable to slow the water's progression through the ground. This is done to allow the plants a long time to utilize more of the nutrition that passes by. One way of slowing the water down is to add organic material that sucks up water without causing any problems. Another way is through mulching, which reduces evaporation. In addition, mulching protects the soil from erosion due to winds.

Mineral composition

The mineral composition in the soil determines how fast the water moves. Pure clay gives the slowest flow of water, while sand provides the fastest. The sweet spot is a mixture of both types so that the soil retains moisture without becoming waterlogged.

Gases

Gases are not usually what you think of when it comes to good versus bad soil. Nevertheless, these are crucial if you want to succeed at growing things. The three primary gases in healthy soil are oxygen, nitrogen and carbon dioxide. Plants use carbon dioxide for photosynthesis, while the oxygen that forms from it is used by creatures living in the soil. Nitrogen is needed for plants to be able to build proteins, to perform photosynthesis, and to form DNA. A balanced exchange of gases between soil and air ensures that there is the right amount of oxygen in the ground – so that plants and micro-organisms can thrive. Without a balanced exchange, there is a risk of potentially harmful levels of other gases building up, such as sulfur dioxide.

The efficiency of the exchange of gases is linked to the structure and composition of the soil and primarily occurs as follows: When it rains (or when irrigation is used), the water penetrates the ground, and the gases are pushed up into the air. The opposite happens when water from within the ground evaporates, and air is drawn into the soil. The most common form of gas exchange, however, depends on the respective pressure of the gases. For example; if the pressure of carbon dioxide is higher in the air than in the ground, the carbon dioxide will move down into the ground to create equilibrium, and vice versa.

Organic matter

Organic matter is the magic ingredient of healthy soil. It provides nutrients, helps with mineral composition, and affects moisture retention. Organic matter also provides food for all the microorganisms living within the soil. You can never have too much organic matter in the ground, and it is important to always return the nutrition that has been removed. This is best done through mulching or composting. When harvesting, it is advisable to leave what you don't need on the ground – such as fibrous waste, leaves, stems, and other organic matter.

Microorganisms

When the gases and water levels in the ground are in balance, and if there is enough organic matter there, you will have a soil teeming with microorganisms. It is these organisms – bacteria, fungi, and insects – that ultimately make for a thriving soil. The microbes help to break down the nutrients present in the soil so that the plants can access them. These organisms also break up the soil, which provides room for roots, air, and water. And, when they eventually die, they become part of the organic matter themselves. The more microorganisms – and the greater variety of species – the better, because the ecosystem then becomes self-regulatory, while also preventing attacks from pests and vermin.

How plants get their nutrition

Everything comes back to the ecosystem around the roots of a plant. One could say that a plant's root system consists of primary taproots that "drink", and secondary fibrous roots that "eat" the nutrition needed in a proper plant diet.

The plants taproots "drink" water and water-soluble nutrients, which in today's agricultural industry almost exclusively comes from the chemical fertilizer NPK that consists mainly of salts where N stands for nitrogen which makes the plant green and lush, P for phosphorus which promotes growth, and K for potassium which is good for blossoms and increased winter hardiness.

Even in organic farming, plants "drink" a saline solution as plants can only absorb nitrogen as nitrates. The difference is that here, the nutrition comes from organic matter, such as manure or blood meal.

However; just "drinking" does not a balanced diet make. The plant also needs to "eat", and the most important factors for this is the work of the microorganisms and the mycorrhiza.

The plant's fibrous roots "eat" the nutrients in the soil with the help of countless microorganisms found on the surface of the fibrous roots. Through these microorganisms, the plant gains access to the soil's nutrients. In return, the soil's microorganisms get their "food" in the form of carbohydrates (starch from photosynthesis) from the plant. A good example is Lolium, also known as ryegrass, which can have 120,000 meters of fibrous roots! In a well functioning ecosystem, the plants and the soil consequently feed each other.

But different types of soils are not equally fertile. This means that not all nutrients are found everywhere, and it is here that mycorrhiza enters the picture.

Mycorrhizal network

The nutritional interplay between some naturally occurring fungal hyphae (mycelium) and the roots of the plants is called mycorrhiza.

You can view mycorrhiza as an additional, giant fishing net for the benefit of both plants and fungi. The fungi's hypha gives the plants a much larger area to collect nutrition from, and the fungi, which cannot bind carbon from the air through photosynthesis, receive their carbon from the plants.

And not only that: One mycelium can be in contact with several plants, which means that the carbon bound in one plant may benefit another. In this, sciophyte is a perfect example. The shadow they live in causes these plants to have limited opportunities to bind carbon from the air themselves, and so the mycorrhiza functions as an extended fishing net to gather carbon from other plants.

In addition to this, mycorrhiza provides several other important factors. The plants get better protection from diseases, they get access to nutrients that the plants themselves can not absorb, and they can withstand significantly lower water levels in the soil without wilting. Mycorrhiza also allows plants to absorb micronutrients directly from minerals through microbiological weathering. Furthermore, the plants can better cope with acidification.

A living and prosperous soil has an airy and crumbly texture that worms love. The more worms there are in the soil, the better it is. In days gone by, the value of soil was actually estimated in relation to the number of worms in it!

All in all, this provides a greater understanding of why the Earth's micro-life is of vital importance. Even if you do not understand everything, it is enough to understand that the industrial agriculture of today, with all its chemical fertilizers, poisons and irrigation systems, gradually destroys, and eventually kills our fertile soils.

The soil's essential nutrients

For the plants to grow and feel good, they need healthy food in the form of different nutrients. There are 14 vital substances, and these are usually divided into two groups.

The substances that the plants need a lot of are called macronutrients – which are nitrogen, potassium, phosphorus, sulfur, magnesium, and calcium. The elements that the plants need smaller amounts of are called micronutrients – and these are iron, manganese, boron, zinc, copper, chlorine, nickel and molybdenum.

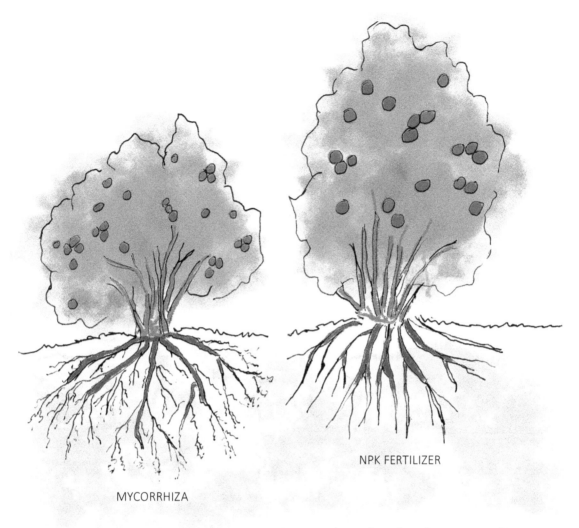

Root systems with and without mycorrhiza

The availability of these substances varies from place to place, as one might guess. That is why nature looks different in different locations. For each local ecosystem, nature sustains itself with the aid of what is available in that place, in the form of decomposed organic matter and the minerals that the plants pull out of the ground.

Essential substances for us people

The interest in micronutrients has increased in both the study of plant nutrition and of human medicine in recent decades. Many things point to the fact that the food grown on soils with insufficient access to micronutrients may contain inadequate levels of certain substances for humans, even if the crop does not show any signs of deficiency itself.

The soil's pH plays a major role

That the pH of the soil should be close to 7 is questioned more and more often. According to newer recommendations, the soil should be more towards the acidic side. In higher pH levels, the risk of a lack of micronutrients increases. In the case of other useful soil organisms, – such as earthworms – recent studies indicate that the availability of fresh organic matter is much more important than the pH level.

High pH

The risk of a deficiency of iron, manganese, boron, zinc, copper, cobalt and nickel is increased by calcination and with increasing pH levels. Many macronutrients are also adversely affected by calcification and a high pH. The risk of nitrogen losses through leaching, denitrification and the loss of ammonia rises with increasing pH. It is also documented that calcification and pH above 6.0 increase the leaching of sulphur, thus leading to a greater need for sulphur fertilization. Phosphor rapidly becomes less accessible for plants when the pH exceeds 6.0. This is because the form of phosphor that plants absorb transforms into a form that is much more difficult for the plants to absorb. The release of the phosphor bound in the ground is more effective at a low pH and is inhibited by calcium.

Raising the pH level is easy; lowering it is significantly harder. Potent pH-reducing organic fertilizers do not exist. Blood meal has a slight pH-lowering effect, and pure sulphur (Flowers of sulphur) lowers the pH as well as adding sulphur. Leguminous plants lower the pH clos-est to the roots, and can thus increase the availability of phosphor and micronutrients in the soil.

Low pH

At low pH levels (usually below 5), the levels of manganese and aluminium in the soil can be so high that it damages the plants. The risk of that is much higher when using chemical fertilizers than when using organic fertilizers. Studies in northern Sweden have shown that high levels of accessible manganese in the soil solution seem to inhibit the absorption of cadmium in the plants. Many studies have shown that mineral fertilizers increase aluminium damage and increase the absorption of cadmium in the crop. Organic material, on the other hand, often decreases the absorption of cadmium.

Location decides the climate

The climate affects us wherever we live, and the place we call home relies on several climate factors. This is something most of us probably learned in school. Still, from a horticultural perspective, it may be useful to brush up on that old knowledge. Therefore, I am going to go through the fundamental factors here.

Our place on Earth

First, an explanation of terms describing where on the globe a place is located.

Latitude

Latitude indicates in degrees how far from the equator a location is. At the equator, the latitude is zero degrees, and at the poles 90 degrees. All latitudes are parallel to each other, and the distance between each latitude is approximately 111 kilometres. On the other hand, the circumference of each latitude decreases as it gets closer to the poles.

Longitud

Longitude, or meridian, is the lateral distance from the so-called prime meridian in Greenwich. Longitude is indicated as positive up to 180 degrees east of Greenwich, and negative to 180 degrees west of Greenwich. Longitudes, unlike latitudes, are not parallel to each other. On the other hand, the length is always the same, since the longitudes go from pole to pole. The length is approximately 20,000 kilometres.

An interesting effect of the Earth's rotation and the 24 hours of the day is that the Earth rotates exactly 15

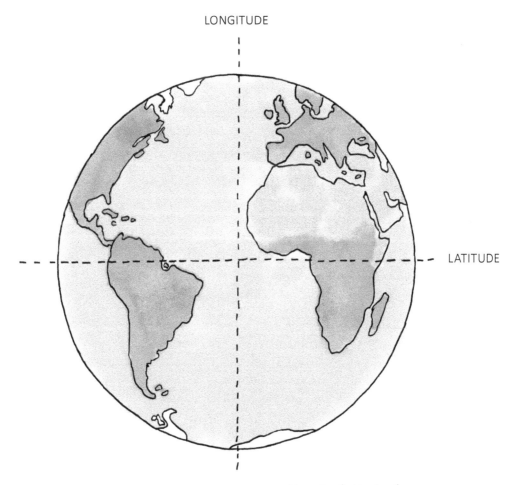

Longitude/Latitude

degrees in an hour regardless of latitude. This means that the Earth's surface spins fastest at the equator and as the slowest at the poles.

Orographic effects

Orography is a study dealing with the formation and features of mountains, as well as highlands, lowlands, valleys, and plains. The orographic effects have a significant impact on the local climate.

Here are some basic terms to know:

Altitude

The climate is affected by how high above the sea level a place is located. The effect is that for every 100 meters - either up or down - from the sea level, you will be one latitude up or down as seen from the equator. Think of it like this:

The higher you get the further north you will come in terms of climate. However, the angle of the sun is not affected, which means that the length of the day is the same, regardless of how high you are from the sea level.

This, in turn, means that even though the climate of the site allows for certain plants, these plants must still be able to cope with the length of the days. A good example is that in the tropical areas we cannot grow our northern plants even though the site is well above the sea level, in a climate similar to our own.

Maritime effect

Oceans, seas, lakes, and even small ponds affect the climate due to the fact that the change in temperatures in water is so much slower than in air. This is especially noticeable on a landscape level. The closer to the sea or a lake, a location is, the less powerful the temperature fluctuations. The opposite is also true – the farther inland, the greater the fluctuations.

That is why the coastal climate is milder and softer in its "mood swings", and the climate farther inland can offer rapid changes in weather, with frigid winters and hot summers.

Rainshadow

In mountainous terrain, the influence of the mountains on the climate is considerable. Depending on the weather, a rain area can either be lifted over the mountains or come to a halt before them.

Both situations mean that a so-called rain shadow is created on the side of the mountain, which isn't reached by the rain area. In this, local knowledge is especially crucial to find out which areas have a weather pattern that repeatedly causes the same side of the mountain to have a rain shadow. That side may have a considerable lack of water compared to surrounding regions.

A location's appearance creates microclimates

Microclimate means the climate within a limited area: Where the microclimate can differ radically from the surrounding environment. The differences are due to the small-scale design of the landscape, the groundwater content, and the design of the vegetation. The microclimate has a significant impact on both the plant- and soil processes, so it is essential to pay attention to all the small details.

The trees show this very clearly. On the side where the sun hits, the lichens rule. Here the temperature and humidity vary considerably. On the shadowy side of the tree, different types of mosses dominate, and the climate is generally both cooler and moister than on the sunny side. The most apparent effect is that trees begin to bloom on the top of the south side and ends on the bottom of the north side. For maple trees, the difference can be up to a total of 10 days.

The ground surface's climate is very different, depending on whether it is covered with plants or not. The plants effectively dampen the highs and lows in the temperature and raise the humidity in the air closest to the ground surface. The naked Earth has no such protection and suffers from substantial temperature changes between day and night. This is especially true for dry soils.

Topography also has a significant impact on microclimate. Especially the colder temperatures are affected by small topographical differences, as cold air is denser than hot air.

In a landscape, this means that the cold air falls into all the valleys and dips – even the smallest ones; which in turn makes these places more prone to frost. The same thing applies along a slope where the cold air slides downwards, only to settle against the first obstacle it meets. This means, for example, that the edge zone between the end of a slope and a grove is more likely to be hit by frost.

How permaculture works

As I wrote in the introduction of the book, permaculture can be a solution to many of our problems – all over the globe. In this chapter, I will briefly explain what it means, first and foremost from a horticultural perspective. The rest of the book will offer information both on the practical aspect and a more in-depth look at how to use permaculture alongside other tools.

Permanent Agriculture

The term permaculture comes from the words of Permanent Agriculture. It was coined in 1978 by Bill Mollison and David Holmgren, both from Australia.

Bill was a researcher and biologist, and as such, he was fascinated by the forest and how it worked when it was allowed to grow freely without human influence. At the same time, he was very committed to finding a solution to the catastrophe he believed the world was heading towards. Despite both involvement and participation in many protests, he felt that nothing he did yielded any visible results.

In the end, he realized that the change that was needed did not come from above, but from below. Therefore, he returned home and began to cultivate crops in ways inspired by the forests and by the indigenous people that remain in the world. From this, his book "Permaculture: A Designers' Manual" was born. A complete manual for everything related to permaculture.

The ethical principles of permaculture

Permaculture is a planning tool that can help us find our way back to a sustainable system – both for us, and the planet as a whole. Permaculture is a holistic approach that can be summarized in the following three ethical principles.

Caring for the Earth
Caring for the Earth means that we recapture the understanding that we are part of the planet and that we live in synergy with all other life.

Caring for people
Caring for people is about finding our way back to a more close-knit community, where people have time for each other – that we can talk, listen, see, and be seen.

Fair distribution
Earth's resources are finite, so we need to manage and fairly distribute these resources in a way so that we do not consume more than what is renewed.

Is change hard?

The answer is both yes and no. The difficulty lies in that the problems are global, and that makes the decision process extremely slow and convoluted. It is the economic doctrine of constant growth in particular which is the driving force behind most of the problems. Unfortunately, the likelihood of a quick change is minimal – because nobody, neither politicians nor financiers want to be the first to take a step towards change. Nobody dares to be the first to put the nail in the coffin on today's shining economy.

How, then, do we stop the runaway train? Here, I believe the saying "if you want something done, do it yourself" comes in handy even if it can sound near impossible. But imagine – what if we can?

Permaculture is about taking the reins ourselves and changing our own way of life. We can do so much more than we think, and a good start is to change our perspective from "something must be done" to "what can I do?".

Because we decide our own actions, as long as we stay within the limits of the law. No one can prevent us from starting – right now – to see things from another perspective. Everything does not need changing all at once; we can take it one step at a time. Because how do you move a mountain? Why; by moving one rock at a time of course.

Therefore, let us start by growing more of our food ourselves. If we don't have that option, we can at least choose sustainable, organic and locally grown produce. If we take a step back and rethink, while asking ourselves "Do I really need this?" we can reduce our consumption while saving both resources and money. As soon as we succeed in taking a step towards a more thoughtful existence, we need less money, which has the added bonus of needing to "work" less. This, in turn, makes the puzzle of life easier to solve.

The 12 design tools of permaculture

In addition to the ethical principles, permaculture has 12 design tools. Within these, many of the permaculture keywords words such as observation, planning, natural balance, resource conservation, renewal, and overall return are repeated throughout. Together, the design tools reflect the three ethical principles of permaculture.

→ *Observe and interact*

→ *Catch and store energy*

→ *Obtain a yield*

→ *Apply self-regulation and accept feedback*

→ *Use and value renewable resources and services*

→ *Produce no waste*

→ *Design from patterns to details*

→ *Integrate rather than segregate*

→ *Use small and slow solutions*

→ *Use and value diversity*

→ *Use edges and value the marginal*

→ *Creatively use and respond to change*

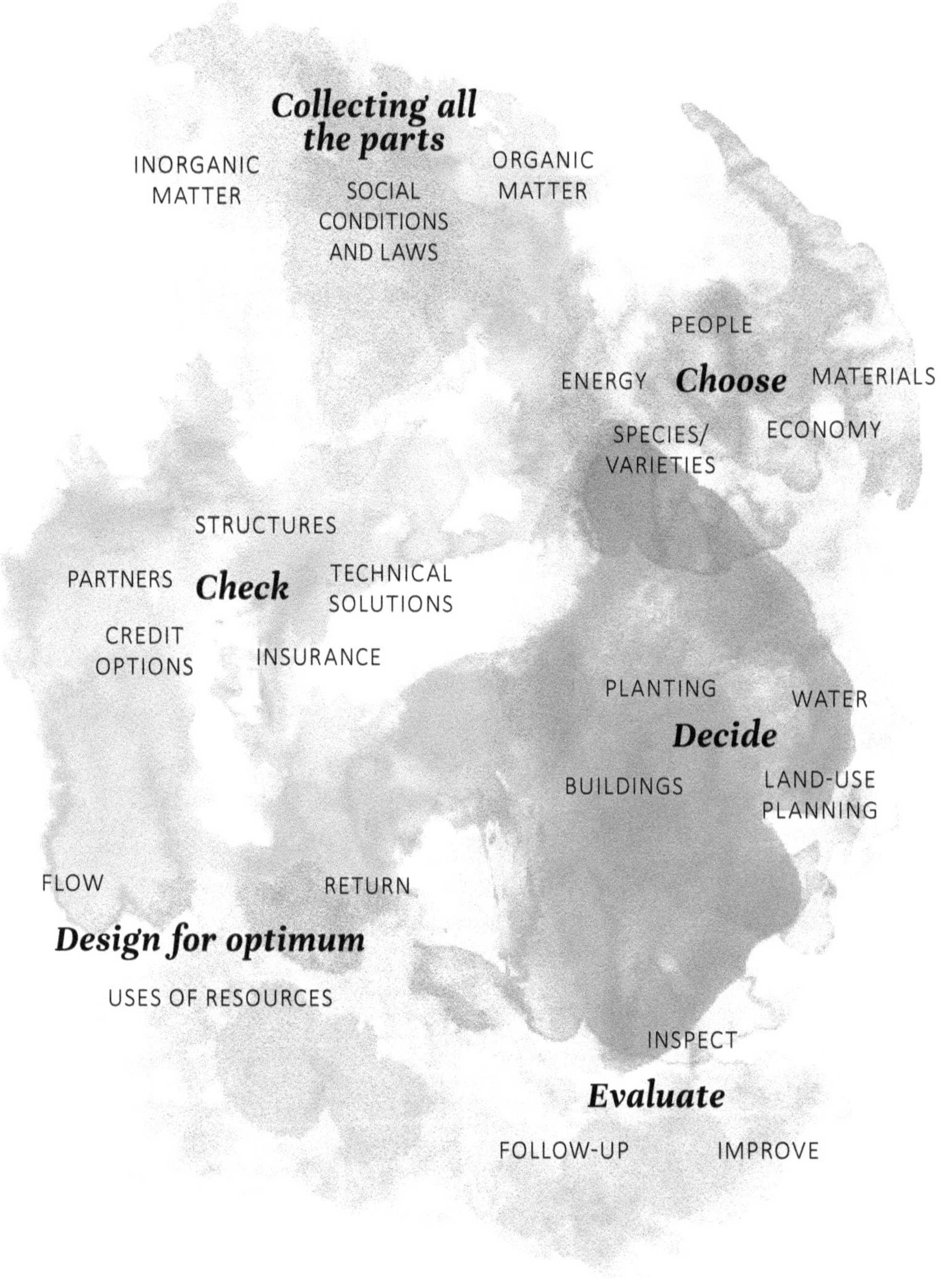

Collecting all the parts

INORGANIC MATTER

SOCIAL CONDITIONS AND LAWS

ORGANIC MATTER

PEOPLE

ENERGY **Choose** MATERIALS

SPECIES/ VARIETIES ECONOMY

STRUCTURES

PARTNERS **Check** TECHNICAL SOLUTIONS

CREDIT OPTIONS INSURANCE

PLANTING WATER

Decide

BUILDINGS LAND-USE PLANNING

FLOW RETURN

Design for optimum

USES OF RESOURCES

INSPECT

Evaluate

FOLLOW-UP IMPROVE

Observe and interact

With increased attention, a little patience, and a sense of curiosity, we will soon learn how nature's systems function. Then we can begin to interact with nature. Because when nature takes care of itself – it performs at its best. It has itself built up complex and regenerative systems under 4.5 billion years, and one might wonder why we humans try to change this.

Order and disorder

Man would like to have things neat and tidy in different contexts, both when growing produce and keeping animals. But our order is not always the same as the one nature wants. In this, the ancient laws of nature apply. These laws require us to get better at understanding and mimicking.

When nature thinks things are in disorder, it results in chaos in the ecological system, and it takes more energy to do anything. Put another way – when creating human order, it can take much more energy to do something than when nature itself manages the order. A good example of chaos in nature is the so-called monocultures. One example is when growing a single crop on large areas for several consecutive years. Another example is when keeping animals in the same pasture year after year, while food and water get transported in.

A rule of thumb that is good to remember is that nature is in disorder when it consumes more energy than what it yields. On the other hand, when a natural order is allowed, nature yields at least as much energy as one put into it.

This may seem a bit confusing, but let us use lawns as an example. To keep lawns green and lush, a lot of energy is used – both diesel and human power – without getting anything back. Thus, clean resource consumption. Here, a small change can make a world of difference. Because if you collect all the grass clippings and use it for covering vegetables, the lawn becomes a useful producer of plant nutrition.

STRUCTURES/BUILDINGS

ENERGY SOURCES

FLOOD RISK

CARDINAL DIRECTION

SOIL EROSION RISK

VEGETATION

FIRE RISK

SUN/SHADE

To observe

ROADS

WATER

SLOPE

WIND

ALTITUDE

CLIMATE LATITUDE

PERCIPITATION

OCEAN/SEA/LAKE DISTANCE

SOIL TYPES

ANIMAL HUSBANDRY

Stressed out and in harmony

The difference between an ecosystem that is stressed and one that is in balance is profound. Stress in nature is defined either as different natural functions being prevented from working, or that some functions are forced to work. Balance is when all the different natural functions cooperate in the best possible way.

When stressing a system, long-term issues are built up. Not only are you removing parts from the system, you are removing the diversity, and also the contacts between the different parts of the system itself. This makes for an empty system that is difficult to restore in the short term.

Nature itself tries to fix these "holes" quickly. Still, often the repair elements are not those that were there from the beginning, and from a human perspective – not even desirable. Like weeds. When a soil system is disturbed, nature tries to fix the problem by favouring certain plants; however, these plants are perceived by us as weeds, that must be combated.

In modern egg production, the chickens that are in a cage or free-standing indoors are an example of a dysfunctional system. In these environments, the hen's only function is to lay eggs, where vast resources are needed to keep the system running. The food needs to be transported there, and the chicken droppings become waste, waste that needs to be dealt with. In such a system, the hen is losing many parts of its natural behaviour, which creates stress and poorer health in the hens, and even poorer egg quality.

After all, when hens live naturally, they give us so much more. They lay their eggs, scratch and process the upper soil layer, eat insects – which causes fewer pests, and last but not least, they fertilize the soil with their droppings. From such a natural system, there will be no waste. The hens are much healthier, and the eggs are not only tastier but also more nutritious.

Catch and store energy

Economizing the use of all types of resources is a phrase heard throughout all form of permaculture. As a whole, this means designing and planning to save and manage the various resources available for each specific situation. When it comes to sunlight, soil, water, fire, and air; it's essential to plan how to best catch and store the energy as long as possible within a system.

The sun is our primary source of energy, and we need to be much better at using it. This can be done, for example, by using solar energy for both heat and electricity, and by designing gardens and buildings to be energy efficient.

Water is becoming increasingly important, as the climate changes and we humans continue to multiply.

Here, a change must come about immediately, so that we use less water. At the same time, more attention needs to be paid to collecting and storing water, as well as creating systems for efficient use, purification, and reuse of water.

Fertile soil is the foundation of everything. If the amount of energy consumed is larger than the amount returned to the soil, the soil will gradually become depleted. In order to nurture the soil, organic farming, composting, mulching, perennial plants, and forest gardens are used. In favourable conditions, geothermal heat can also be extracted.

The basic rule for air is simple; you do not pollute it further. The natural processes within green plants mean that carbon dioxide is bound in the soil, new oxygen is formed, and the air is purified. This means that the more you grow, the better the air will be – provided that the crops are free from pollutants. As an energy source, the air temperature can be utilized by means of an air source heat pump.

The wind needs to be handled in a well-thought-out manner, so that you get the biggest benefit, with the smallest amount of damage. Just think about how much a storm can destroy – or how an especially windy place in your yard could provide energy using a small windmill.

Fire is an element that we do not have such big problems with so far up here in the north, but with climate change, the risks are increasing. Therefore, it may be useful to also consider the site's risk of fire and how this can be prevented and managed if something does happen.

Last but not least, taking into account human energy consumption is vital. Sometimes we do this without thinking, like planting the spices as close to the kitchen as possible. While other things may end up being too far away, and end up being hard to manage. Within permaculture, this has evolved into an encompassing mindset, a tool that divides the plot into different zones and sectors to optimize energy consumption.

At the same time, it is important not to store more energy than what might be necessary; because even energy can become a form of waste. Energy waste occurs, for example, in the rushed production of both crops and animals, where large amounts of energy are forced into the system, to provide the highest possible return. This creates large amounts of excess energy that often causes consequential issues, such as chemical fertilizers leakages into nature, and manure that goes to waste.

Obtain a yield

When you put work, time, and money into something, you want that to bear fruit, but an ecosystem in balance does not provide a surplus. In order to get a surplus – such as a bountiful harvest of vegetables – it is important to add just the right amount of additional energy the harvest requires. After harvest, the lost energy must be returned; otherwise, the soil will be depleted over time.

Types of yields
Today, almost everything is about product yield – and above all, economic returns. However, several other forms of yields are important to the whole. In a permaculture system, all the different types of yields are seen to, while also avoiding the overuse of resources.

Product yield
Product yield is the surplus in any part of the system.

System yield
System yield is the sum of the surplus energy produced, preserved, stored, reused, or converted by the system in addition to the system's own growth, reproduction, and maintenance needs.

General yield
General yield means things such as health, nutrition, security, a satisfying social life, lifestyle, and freedom. An example of a bad general yield return is a high-rise area that has a sole purpose of generating as much money as possible, on as small a surface as possible. Little or no consideration is given to other needs.

Apply self-regulation and accept feedback
We need to start thinking about the long run again. Driving on in the wheel tracks of today's short-term perspective is not providing new results. With the help of permaculture, we can create self-regulating systems that eliminate much of today's overconsumption.

Everything done in today's society – with very few exceptions – is incredibly short-lived. The most important thing is to make a profit, and the faster everything spins, the shorter the time you think you have. Most have a perspective spanning days and weeks; plans for the week, plans for the weekend, plans for holiday leave, and so on. We rarely plan what to do for next summer, next Christmas, the next five years...

Almost no functions in society are based on long-term thinking. Why; even some laws only apply up until the next election. And looking at the short-term view, the economy and politic climate seems to be running on quick actions.

Sometimes, it is said that it was better in the old days, which is a truth with modification – but in terms of personal responsibility and long-term thinking, it really was better. The adult generation thought about the future of their children and grandchildren in a completely different way than they do today.

To think like the saying of the native Americans'. "In every deliberation, we must consider the impact on the seventh generation" may feel unreasonable. But, on the other hand, if we can start thinking about the grandchildren in our decisions, much will be won. How will their world look if we continue as we are now? It is actually only 60 years since the more serious dark side of industrialization accelerated. In these 60 years, we have already destroyed a lot. It is, therefore, high time that we begin to restore the natural balance.

What this has to do with permaculture may not be obvious, but the systems thinking of permaculture works very well as a solution to the problem. With systems thinking, our ecosystem does not hurt, it does not take out more than the system allows, and no resources are drained. Finally, we get a sustainable system that does not use Earth's resources – including us humans – to a greater extent than what they need to be recreated.

Use and value renewable resources and services
Within permaculture, resources are used according to the "enough" principle to create sustainable resource usage. This means producing enough to satisfy the real needs, not producing what you think but don't need, and, above all, not producing a surplus just because you can.

Permaculture also applies to the rule of giving back. This means that whatever you take from nature, you have to give back – at least as much, or more, than what you have taken. It means sustainable systems, where the renewable source is truly renewed at the same rate as it is used.

The problem is that the concept "sustainable" has become a marketing ploy, where different industries increasingly emphasize their products as both sustainable and renewable. But they rarely, or even very rarely explain how the renewal happens. They often talk about how good it is to buy that which is renewable.

Unfortunately, the companies that sell these renewable products are rarely responsible for the renewal itself. It is, therefore, hard to know what is really sustainable.

Produce no waste
In the wild, there is no such thing as waste – everything is reused. A natural ecosystem has an energy balance, in which all living organisms retain their complex forms and functions through a continuous exchange of energy with their environment.

A surplus or deficit in the environment that nature can't deal with creates, as I said before, imbalance and disorder. The natural system is negatively affected by imbalance, and selected parts – or even the entire system – crashes.

That is why every part of permaculture is essential. This applies to everything from the observation at the beginning to the creation of a design and an ecosystem that provides that "enough" surplus for immediate use, and what you need for later, or to sell or give away. Again, a surplus that is not used is waste and overuse.

Design from patterns to details

Design means shaping and placing all the parts in a working whole; an ecosystem. To do that, you need to find out what different elements, functions, zones, and sectors are available. We also need to understand how the different parts are connected and which natural patterns are present, all the while creating a picture of the whole for yourself.

Different elements

When designing, it is primarily these elements that you work with:

→ *People*

→ *Plants*

→ *Animals*

→ *Water*

→ *Soil*

→ *Structures such as houses, corridors, garages, and roads*

Different features

Each element then has several different features that one must take into account:

→ *Needs*

→ *Behaviour*

→ *Products*

→ *Neighbourhood character*

Different contacts

Different elements and features have a variety of contact routes:

→ *What does an element fit best with?*

→ *How do the contact routes look?*

→ *How many contact routes are there?*

→ *Is there any contact routes that connect more than two elements?*

Location orientation

Cardinal points, slopes, and other elements casting shade determine how much sun you can get, which is especially noticeable the farther north you get.

Most often, it is said that a southerly position is the best orientation. Still, optimally, the sweet spot is a position that is slightly twisted to the southwest, because then the last warming rays of the evening can be utilized. It is, therefore, more challenging to grow on a slope turning away from the sun, than on one facing south or southwest.

How steep of a slope also determines the growing possibilities, where 18 degrees is said to be the limit of what is reasonable to grow on. On the other hand, a slight slope can do a lot of the work for us if we are taking advantage of gravity.

Human energy

Human energy is the central energy that has to suffice and reach the entire system. The more factors that can relieve us at work, the more efficiently you can manage your energy. Here, different animals play an essential role in everything from preparing the ground, to pulling carts and carrying things.

Patterns

Everything primarily consists of patterns. From the intricate pattern on the petals of the smallest flower to the appearance of the old natural forest – to the whole Earth's weather system, and out into the universe. Even so, there are not that many different types of patterns – but of each pattern, there is an infinite number of variants.

Common patterns

Fingerprints consist of either round or open arcs. There is no fingerprint identical to another. No matter if you look at those that have been, those that are, or those that will be.

Waves of all kinds are harmonic patterns that have infinite variations of both size and speed.

Hexagonal patterns are for example the basic pattern for both the bee's honeycombs, and for snowflakes. The mind baffles when you think that no snowflake is the same as another— either of those who have fallen or of those who are going to fall. And not even these hexagonal patterns have to be symmetrical.

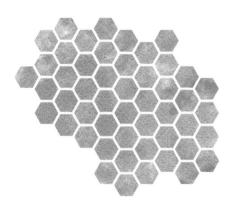

Branch patterns are when a trunk splits up into three branches into nine, into 27, into 81 – all while lessening in size, but growing greater in numbers and in the speed of growth. Examples of branch patterns are seen not only in trees, but also in rivers, and in our own lungs. The branches neither are or need to be geometric. Just think of how different trees look, and yet they all have the same basic pattern.

The natural forest is also shaped according to natural patterns, where everything is adapted for the forest to grow and thrive in the best way possible. This pattern of knowledge is used, for example, when designing a forest garden.

Even our own memories are shaped in patterns. Think about how we, to remember something important, link this information with some other logical memory, just to make sure that the information is not lost.

In the old days; before writing and counting, knowledge was transmitted through patterns in songs, rhymes, artwork, symbols, etcetera. The importance of this knowledge transfer is clearly apparent in the Polynesian people's approximately 2000 navigation

songs that they used at sea. In addition, these songs were often linked to the stars, the appearance of the water, and the movement patterns of the waves. With the built-in knowledge of these songs, the Polynesians made their way to both South America and New Zealand over rough and foreign seas.

Sectors

A landscape, whether large or small, consists of different sectors with different conditions. Within permaculture, knowledge of these sectors and their conditions are important components of the design stage. The most common sectors you work with are:

→ *Sun*

→ *Wind*

→ *Views*

→ *Noise*

→ *Dust*

→ *Odour*

→ *Flood risks*

Zones

To effectively plan energy consumption, an area is divided into five different zones. On a plot, the home is zone zero, the area closest to the house is zone one, and zone two, three, four and five are then located outwards in turn, as seen from the house. An example of Zone Five is an adjacent forest that you only visit a few times a year.

To illustrate the zones, you usually draw them as in the picture, but that does not mean that that is the only right way to go about it. How the zone division looks, and how many of the zones used in an area is determined by the environment and by what you want to achieve.

Comparing this zone division with what I previously told of today's indigenous people, it is easy to see that this is an ancient knowledge that was already in our old traditional societies, with no designer being there and putting in their two cents. For them, it was simply the smartest way.

Integrate rather than segregate

Everywhere in nature, you can see how plants, insects, animals, and bacteria are integrated into well-functioning ecosystems. Ecosystems that are always based on cooperation and support between the components in it. An excellent example of such an ecosystem is the fruit tree. It provides not only fruit, but also shade, and protection

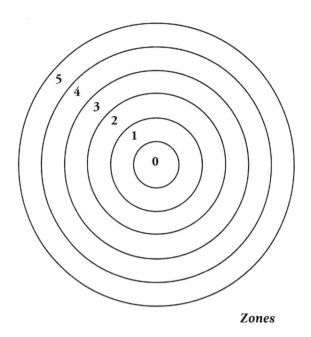

Zones

for both humans and animals. It nourishes insects when it blossoms and is beautiful to look at. The tree draws water from the ground to the benefit of other plants, it may support a climbing plant, and it eventually turns into lumber and firewood.

Within permaculture, plans are made to best imitate these integrated systems. This means that each element of a design should provide multiple features and that several elements support each feature. Difficult to understand? Look at it this way – when growing carrots, it is wise to cultivate several varieties, where some will be ready early in the season, some in the middle, and some late season. If you can then manage to get varieties that have different abilities to cope with different temperatures and precipitates, you have increased the chances of getting a harvest, even if there are problems during the season.

Use small and slow solutions

Small and slow systems are easier to maintain than large and fast ones. Resources are also usually better utilized, and the results are more sustainable in a small-scale system. It goes without saying that it is harder to respond and act in time when a large and fast-moving system threatens to go wrong than when a small and slow system does. For example; an oncoming storm is hitting the next day, and you have to manage to take care of the harvest before the storm comes. In this situation, size plays a decisive role. Wherein a small system, there is a greater chance of saving the entire harvest.

The goal of a design is therefore to create an ideal system that produces enough energy for the system itself to function, as well as providing the surplus needed while also storing resources required to maintain the balance of the system, in the event of uneven energy supplies.

Use and value diversity

Diversity in an ecosystem reduces its vulnerability and increases the possibility of a sustainable system. However – the variety itself isn't enough. The diversity must be built on the mutual relationships of the parts within the system. The parts must fit together in the best way possible – otherwise, variety will lead to more chaos.

The age of the plants is an essential part of this. Because when all the plants in a system are the same age, the plants are more vulnerable to diseases, pests, and extreme weather conditions than in a system where their ages vary. If all the plants in a system are old, the use of energy becomes less effective, as mature plants need more energy to maintain their size, which lowers the yield. The same goes for a system with only young and new plants, where most of the energy is used to grow and establish the plants, rather than producing a yield. In a system with adult plants, the production is at its best – but no one is there to take over when production decreases.

The most efficient ecosystem is one where some of the plants are babies, some are adults that provides the yield, and some are old. A system where the mature plants have grown big provides protection for the babies that can then grow strong to take over while the adults age.

In a permacultural system, this knowledge is used, and you are continually striving towards diversity in as many different ways as possible. This is done through intercropping, varying plant ages, and by mixing annual and perennial plants. Another way to create diversity is to extend the growing season with:

→ *Early, mid-season, and late kinds to reduce the yield loss due to various problems*

→ *Greenhouses*

→ *Perennials for an early spring harvest*

→ *Plants that yield harvests from both leaves, seeds, roots, fruits, and berries as they increase both harvest time and diversity through multiple harvests from the same plant*

It is also important to choose crops that deal well with storage, which can be done through freezing, drying, or refining. This way, you will enjoy the harvest even in the cold of winter.

In theory, the yield from a sustainable system is unlimited, because if you look for really carefully – there is always space for something new.

Here are three types of spaces to look out for:

1. *A space in places, which can be a place to be in, to fit in, and find food in – a place with a roof over your head, space to function in*

2. *A space in time that may be cycles of recurring occasions – such as annual, seasonal, monthly, weekly, and daily*

3. *Space in spacetime can be a schedule that connects... well, space and time.*

A few examples are:

→ *Vertical spaces*

→ *Location*

→ *Different Zones*

→ *Different soil types*

→ *Different depths of water*

→ *Different slopes*

→ *Different forms of flow through a system*

→ *Boundaries between edges and margins*

Use edges and value the marginal

Margins and edges – edge zones – means the area where two different environments meet. When environments that naturally work together meet, any diversity becomes significantly higher than in each individual environment. This is a new type of outlook for many will lead to you looking at the edge zones from a new perspective. You not only benefit the crops, but you also save a lot of time on weeding.

The advantages of planning the edge zones are many. Their diversity favours and supports all the plants within, that then attracts a larger variety of useful insects, which in turn attract birds and other insecticidal animals. The edge zones also provide variation in light, shadow, temperature, and moisture – creating microclimates.

A good example is the edge of a forest, which is the edge zone between forest and field. There are many more advantages here than what the forest and the fields can offer separately. At the edge of a forest, there is sun and shade, nutrition that falls when wind and rain come to a halt against the trees, protection from storms and other natural phenomena, and proximity to food in both forest and field. This makes the edge of a forest an attractive place for both plants and animals. All in all, diversity is created on a bigger scale here than in the forest or in the field.

Today, many of the natural edge zones have disappeared, since the industrialization of agriculture has meant that many small patches of dirt have been merged into large fields. This has led to a sharp decline in diversity in vast agricultural areas, and people are now trying to recreate at least some of these essential edge zones.

Another clear edge zone is the one between ground and water – just like the woods, the water's edge offers many more advantages than what water and ground can offer separately. A final example is an area between plants and rocks, where the stones function as both plant protection and as heat generators, which means that plants that generally cannot live in a specific culture climate can perform well next to a large rock.

Creatively use and respond to change

We adapt in every situation to something that in the next step, hopefully, is better than the last. The change is constant throughout our lives, and without change, we would not survive. It is vital that we can adapt and be resilient – that we can continue to explore opportunities, and choose what fits best, and that we can create visions to see the changes that are going to happen.

The same is true when gardening – since even nature changes and adapts to that which is offered. The knowledge about the process of change is used within permaculture when creating a vision – a design – for a balanced landscape that can grow and expand as work is put into it.

Seeing each place as unique, and using the fundamental rules of the game that nature has taught us, can give amazing results in a seemingly hopeless place. Even though you cannot do everything in certain areas, there is always a solution. Therefore, the answer to most of the issues within permaculture is "It depends." Thus, when you understand what it is dependent on – you will have the rest of the answer. Within permaculture, it is often said that the problem itself is the solution.

The same way of thinking applies when trying to turn limitations into resources. Everything can be a positive resource if you leave behind the old ways of thinking. Take, for example, water in the children's pool, fallen branches, or branches that are left over when trees and shrubs are trimmed, and not to mention all the leaves in autumn. None of this is waste or rubbish, in that everything is perfect to use for something else. The water can be used for those plants that need more water than others, and branches of different varieties can be the start of a new garden bed.

The same thing goes for the leaves – they can be mulching material and thus soil improvement in an already existing bed, or they can, together with the branches, become a new garden bed. When you work with this mindset, it is relatively common to have a lack of this type of waste. Then, you simply have to go to the neighbour and ask for their "rubbish"!

Edge zones of the woddland edge

A contemporary "old-fashioned" village

A village permeated by a permacultural approach looks a lot like an old-fashioned, traditional village. The big difference is that in the modern village - old wisdom and functions have been supplemented with new technology and knowledge.

A sustainable village

To put it simply, a permacultural village is created by working together to produce a vision and a plan for how the land is managed in the long term. A group of talented designers then presents the details along with all the other necessary areas of expertise.

The overall vision is something that the "elders" in the village are responsible for. The area closest to each house is private land, and each family takes responsibility for that. Responsibility for the rest of the land is shared jointly by those who are involved in the village.

An ideal village - even one that follows old traditional patterns - consists of a cluster of ten houses surrounded by a communal area for families, cultivation, and other joint activities and ventures. A city with at least three clusters - that is, 30 families - has enough people theoretically to generate jobs within the village. The income comes from the production and processing of products and services. Members can also start businesses on communal land where the yield is free for those who live in the village, while the surplus can be sold outside of it. Usually, each person has several tasks or jobs in several different areas of the village.

Villages built up in this way have shown that this creates an increased tolerance and respect for both humans and animals. Interestingly, in communities that have been around for a while - there are markedly fewer houses that are sold and, fewer divorces.

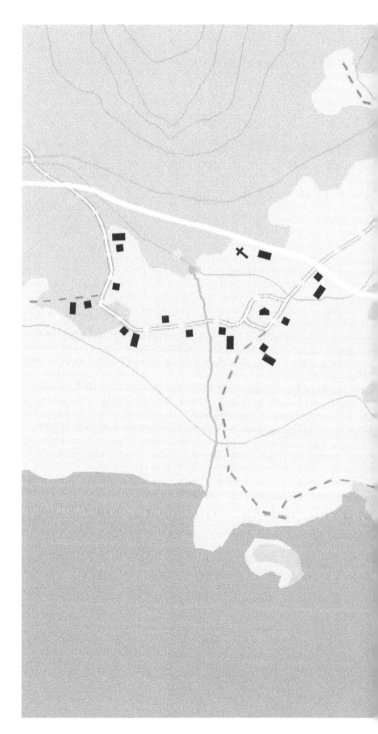

FROM
plans
TO
plants
TO
harvest

WHAT TO DO

Getting to know the place you live on

The first step to using permaculture as a tool for planning gardens and cultivations is to get to know nature again – for real. For many, this can be an entirely new way of looking at both nature and the process of growing things.

Climate change

We have already started to notice that the climate is changing, but we don't truly know how our surroundings will look in the future. To still be prepared for what may come, and to cope with different situations in the best way, permaculture and its tools can come well in hand. Listed below are some possible challenges and how to solve them.

The challenges of our climate

Due to weather changes, new situations arise such as:

→ *Changed seasonal behaviour*

→ *More powerful rainfall combined with colder weather conditions when it comes to rain, hail, and snow*

→ *More and bigger floods*

→ *Erosion of roads and soil*

→ *Heavier temperature fluctuations*

→ *Heavier winds*

→ *Longer dry spells*

→ *Lack of drinking water*

→ *Increased crop failure*

→ *More disease and pest attacks*

→ *Reduced number of pollinators*

→ *Harvest loss when fruit trees and berry bushes are first tricked into flowering too early, only to freeze when temperatures quickly drop again.*

→ *Freezing of plants due to reduced amounts of snow*

→ *Deeper ground frost due to persistent cold*

→ *Increased sea levels*

Proposed solutions to climate challenges

Here are some suggestions for solutions that can increase the ability to handle the changes we face.

→ *To prioritize small, simple, sustainable, and dynamic solutions that are human-friendly, hand-made, and easily accessible*

→ *Creating as diverse solutions as possible in all areas; food supply, water, cultivation, energy, transport, and so on*

→ *To create different buffers and protections with microclimates, for example; greenhouses of various kinds, wind shelters, ponds, stone and rock formations, snow protection on both roofs and in vulnerable places*

→ *To collaborate and build up local networks to gather knowledge and experience on-site*

→ *To increase your own – and the local – self-sufficiency in every possible way, thereby increasing readiness to handle different types of crises and rapid changes*

→ *Finally, it is wise to plan for one zone colder, and two zones warmer*

Observing nature

Indigenous people often say that modern folk does not sit still long enough, or often enough to see and to understand. Unfortunately, it is quite right, but when you start with permaculture, you have to learn the art of being still again.

Observation is fundamental and cannot be replaced by any other knowledge. In the beginning, you need to spend a lot of time observing your plot, both as a whole – and in every detail. To pay attention in different ways to what is happening in different places, what is growing where, how the wind blows, where hot and cold areas are, where the sun and shade hits, and so on. If you are new to a place, it is best to let this observation part take time, preferably a whole year; to experience all seasons. First, you start by just observing and collecting information, not thinking, analyzing, or deciding.

The more time you devote to observation, the more knowledge you acquire about what happens – large and small. Maybe there is a corner that feels downright impossible. Take some extra time to discover this location's conditions with wind, water, sun, soil – and that which is already growing there. By studying nature, what has happened and is happening, it will become easier to find the solution. Over time, so many different events have been gathered that one can, with reasonable probability, be able to predict what will happen in a place in the future. It is then that that impossible space can suddenly become an opportunity.

Compare different environments

With a new-found approach, one can then compare one's own plot with other similar areas in the vicinity. Where weather and precipitation are about the same – since it provides additional knowledge. It can be a plot, a park or a wild nature site that resembles your space, except with other details and solutions that can be helpful in the planning stage.

A pleasant way of learning more is to talk to gardeners in the neighbourhood about how they have designed their garden, what experiences they have in regards to weather, wind, and so on.

To learn from nature's solutions, it is good to go out in the forest or other wild areas. There; you sit down, look around, and think about what you see. From where do the nutrition and water come? Are there different plants in the sun as compared to the shade? Where is it windy, and where is it still? Which animals are there? Then compare all this with the conditions on your plot. After observing for a while, I am sure that you will agree that nature is far wiser than us. Because in this place everything works without anyone coming in and watering, weeding, and generally making a fuss.

Experience nature

Some things cannot be learned simply by seeing – they must be experienced, with all your senses. A place can tell you a lot about what would thrive there, by taking the time to experience the area during different seasons. If you think the place feels warm, cold, windy, snug, or exposed; the plants will likely feel the same. If you get a certain feel for a place, you should trust that feeling and combine it with what you have seen and heard. You will then have a solid foundation for how the place should be used.

Identify patterns

Nature has its own patterns, both large and small. With plenty of time set aside for observation, you will soon discover how different patterns reappear. For example – how particular plants appear together, in which areas the fog likes to settle, where there is water in the spring when the snow melts, or where the soil almost always dries out. All of this gives clues as to how the many parts of nature are connected within this particular place.

Analyze your observations

Analyzing is not as complicated as it sounds. It simply means looking at the different parts from every pos-

sible angle, thinking about what needs to be changed, removed, or added. Maybe you want some animals. How would that work, what is required, and what does it yield? Or, if you're going to grow lots of potatoes. Where is the best place for that, or do you need to change something to make a better place? It is about looking at how a change would work with the surroundings. By thinking through things properly before implementing change, the chances of success are significantly increased. When one discovers that a problem is a solution to another, which in turn solves the next one and so on, then the permaculture penny has dropped. From this, a whole emerges – a functioning ecosystem.

To map and to plan

Making a map of your garden is a must. It does not have to be accurate down to the centimetre. Still, it does need to be proportionate so that you can sketch out different solutions. The easiest way is to make a basic map with all the fixed parts that you cannot or do not want to change, such as buildings, roads, water, and trees. Then you can get a sheet of tracing paper, put it over the basic design and let your imagination flow. If you want to make the maps a little more precise, there are various professional maps available. If it is your land, there is undoubtedly a land record to start with. Then there are the maps on Google and similar sites that also work well for the purpose.

On the underlying map, you need only include that which already exists, not the things you are planning to create. Here is a list of things that ought to be included:

→ *Boundaries, such as land borders, fences, hedges, and walls*

→ *Buildings of all sorts, such as residential buildings, garages, and greenhouses*

→ *Roads, paths, gates, and other things that can affect access to the area*

→ *Height curves – keep in mind that a slope of more than 30 degrees considerably increases the risk of erosion damage*

→ *Existing water*

→ *Existing vegetation: Pay special attention to whether there are any toxic or particularly invasive species to take into account*

→ *The animal situation – if there are any protected species or pests around*

Climate data

If you have lived for a long time on the land, you probably already have a good idea of the climate there; how much it rains, when the first frost comes, and so on. If you do not have that advantage, there are a lot of facts and data on national or local weather websites. It is also possible to find out where and when the sun rises and sets for different places during the season by doing a web search. With more information about the climate itself, it becomes easier to plan for wind protection, water collection, and where you can best create zones with suitable microclimates. Also remember that when designing, you always want to know which extremes you may need to handle; such as the hottest day, the coldest night, the heaviest rain, and the worst blizzard.

Then, make a climate map where important climate facts are marked out, for example:

→ *Soil properties like wet, dry, good soil, mountainous, etcetera*

→ *Cumbersome wind conditions, cold, hot, or cool*

→ *Areas with sun and shade*

→ *Other factors such as large stones, especially frosty areas, flood risks, and other things that can affect the microclimate*

Water supply

As mentioned before, water is vital. Control of the water supply is a must if one is to succeed in creating a sustainable ecosystem. Therefore, one starts by looking for the highest and longest opportunity to collect water. This can involve groundwater or various forms of rainwater collection; how well the collection point matches the consumption points – perhaps it needs to be pumped out – and so on.

In some places, there is plenty of water, but that does not mean you can skip the planning. Given the weather changes, it is difficult to predict what it will look like in 5–10 years. In other words, regardless of how much or how little water there is, a well-functioning water solution is fundamental. The questions are many, as, for example: Is there any natural water that enters the plot? In what ways can this be collected? Does the plot slope so that you can use the layers of gravity to get the water where you want it, or do you need a pump?

Only when it is clear how the water issue will be solved is it time to plan where all the other parts should be placed.

Plant facts

Within permaculture, one always tries to favour the local plants and, in particular, those that are perennial – to increase the plant system's durability. Check with nurs-

eries, garden associations, and talk to neighbours about what is growing well in the area. Who knows, maybe some sort of local plant variety will appear.

The perfect spot

Determining where various things should be put and what the best spot for a thing is, depends on many circumstances – so let it take its time. Here are some common questions to start with:

→ *Where should the gardens be to get the best conditions in regards to water, sun, wind protection, accessibility, and so on?*

→ *What to grow in the garden beds – trees, shrubs, perennials or annuals? Which plants are suitable for the whole?*

→ *How does the ground look in between the beds – is it grassy or woody? Do any trees need to be removed?*

→ *Do you want to have a more extensive traditional garden to grow – for example – a lot of potatoes?*

→ *Where is the best place in that case?*

→ *Do you want to be able to keep chickens or ducks?* → *Where do these fit into the whole?*

→ *Is it too sunny, shady or windy in any place?*

→ *Do these conditions need to be changed, or is any of the problems the solution to something else. Such as solar power or wind power?*

→ *Can the shade be used positively by keeping animals in that area, or perhaps as a comfortable sitting area for hot summer days?*

Take notes, notes, notes – all the time! It makes for good reference material for you to go back to.

Planning policies

In Sweden, there are many rules to take into account, as I am sure there is in your country. There are a lot of regulations regarding buildings, water, and sewage. If you are unsure – especially if you live within a so-called planned area, it is best to contact the municipality.

Desires and visions

In order to gather your thoughts and set the bar for the project, it is good to write down what you want to achieve. Then, ask yourself questions like these – both before the work begins, and throughout the process:

→ *How much time can be spent on it*

→ *How much money and resources is there*

→ *How much experience is available*

→ *What do you want to accomplish*

→ *What do you want to get out of the project*

→ *What do you think about the place*

→ *What do you want to change*

Remember to always take into account the amount of work required – that during construction, much more energy is needed to move something uphill rather than downhill, and to consider the energy consumption for the regular maintenance when everything is finished.

Evaluation

When you think that you have a first draft ready, it is time to make the initial evaluation. Try to see things from every possible angle, and ask for the help of others to see details you might have missed.

Let this process take time. Examine which parts are suitable in which zones and sectors. Which plants thrive to-gether, and which ones hate each other.

One trick I like is to make so-called contact lists. I personally make one list of plants, one with different concepts as on, in, over, under, beside, etcetera, and one with various structures such as dams, fences, greenhouses, and trellises. Then I put the lists next to each other and start linking plants, concepts, and structures. All of a sudden, a lot of new ideas emerge on how the design can be constructed. Here is an example of such a list.

When you get somewhat of a clearer picture of what you need, it is wise to start making simple sketches. This process may feel overwhelming, but as Bill Mollison often says, "You start with your nose, then your hands, your back door, your doorstep." That is to say, take a little bit at a time and start with zone one closest to the house, because it is easier to correct small mistakes rather than big ones. Once you get started, I promise that stopping is a lot harder than continuing!

Implementation

It is always good to have a proper plan for the work ahead – since certain things need to be done in a specific order.

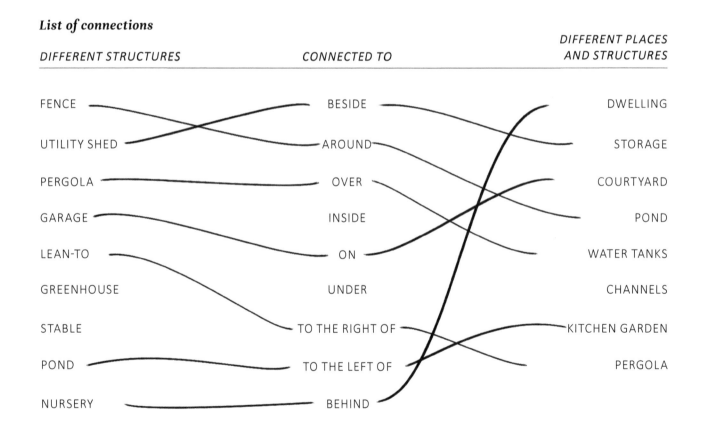

List of connections

DIFFERENT STRUCTURES	CONNECTED TO	DIFFERENT PLACES AND STRUCTURES
FENCE	BESIDE	DWELLING
UTILITY SHED	AROUND	STORAGE
PERGOLA	OVER	COURTYARD
GARAGE	INSIDE	POND
LEAN-TO	ON	WATER TANKS
GREENHOUSE	UNDER	CHANNELS
STABLE	TO THE RIGHT OF	KITCHEN GARDEN
POND	TO THE LEFT OF	PERGOLA
NURSERY	BEHIND	

A wind shelter, for example, must be created before sensitive plants are planted. That all digging and planning work must be done first so that you do not risk destroying anything during it.

Keep in mind that time, energy, and money will last you through the implementation of the design. If necessary, split the plan into several sub-plans, preferably over several years.

Improvements

When the design is completed and a few years have passed, it may be time to go through it to see if it is developing as you intended, or if something needs to be improved. Taking a little at a time, and learning from any mistakes is often very valuable.

The best way to plan a landscape

When you plan a landscape – whether it is a small plot or a farm – you need answers to quite a few questions. Maybe this feels like a lot to keep track of in the beginning, but if you take it a little bit at a time, most of it will turn out to be fairly obvious. To make it easy to grasp, I am sharing a finished landscape design.

Questions to start with

To get into the right mindset, you can start by answering the following questions:

→ *How are your taps supplied with drinking water*

→ *Where do the household's energy and electricity come from*

→ *What happens to your waste*

→ *What was the total amount of rainfall in your area last year*

→ *When was the last storm or flood*

→ *In what place do you get the most wind; wintertime, summertime*

→ *How long is the growing season where you live*

→ *Are there any edible wild plants in your area*

→ *What wild animals live in your area*

→ *Is there a risk of wildlife damage*

→ *What are the most common tree varieties in your area*

→ *Which birds thrive on the site; in winter, summer, year-round*

→ *How did the landscape look like; 100 years ago, 50 years ago*

→ *How did people live in your area in the past?*

→ *Which is the highest and lowest solar angle, respectively*

→ *Which direction is south*

→ *Which wildflowers come first in the spring*

→ *Do the first flowers provide food for pollinators*

Ängsvägen – An example

Welcome to Ängsvägen, home to two adults and two children. The name means Meadow Road, which is a fitting name.

The current situation and the owners' wishes for the land

The family that lives here wants to, in the long run, become more self-sufficient. Today, both work full time, but they plan to cut down on working hours within a few years. Right now, only a small part of the plot is used, where the most used parts are the wooden decks around the house, a smaller cultivation area just below the house, and a small potato field on the meadow down below.

The house is a converted summer house and has electric heating. Still, thanks to an efficient fireplace and a newly installed air heat pump, they have already considerably lowered their heating costs. To save even more money, they want a greenhouse near the entrance, on the southwest side of the house. There is also an outdoor fireplace that they want to build the greenhouse around, to extend the growing season. The idea is to be able to heat the area during periods of frost. Between the greenhouse and the entrance, they want a pergola – to provide extra shade in the summer. To further reduce the temperature indoors during hot summer periods, they want a so-called solar chimney installed. Further down the road, they also want to install solar panels for hot water in the summer.

The plot is mostly mountainous, partly with steep sections. The northern and northwestern part above the house consists of 90 per cent of rocky cliffs, with only small cultivable areas for – for example – herbs and other drought-resistant plants. The areas below the house; the middle part of the plot, has a smaller terraced area which today is used for a small vegetable garden. This area they want to expand and plan better. The rest of the middle part of the plot the family wants to distribute so that the upper part becomes a small "food forest" with fruit trees and berry bushes. On the rest of it, they want space for both chickens and ducks. Down in the meadow, they want to grow most of their vegetables in raised beds – both to promote the microclimate, and to get a more comfortable working height.

The best area for gardening is the flat meadow at the bottom of the plot's southeastern area. This is where they want to grow most of their vegetables – partly for immediate consumption, but above all to save for the long winters. The family understands and accepts that there will be some extra effort due to having the crop and garden area further away from the house. Vegetables for daily use will, therefore, be grown in the smaller garden area that is closer to the house.

Water supply consists of tap water and rainwater. The tap water comes from its own well, which unfortunately is not very tasty; however, it is serviceable. If they need water during, for example, a longer power outage, there is a communal hand pump closeby. Also, only 300 meters away is a smaller lake that can be used as an extra water supply if needed. However, drinking water must be collected a few kilometres away.

For better growing conditions, they want a water system that starts at the top of the forest garden. The system shall consist of water collecting trenches, so-called swales, with built-up garden beds along the bottom of the swale. The water will then – through controlled overflows – continue down past a duck pond and further down to a larger pond on the western side of the meadow, which will serve

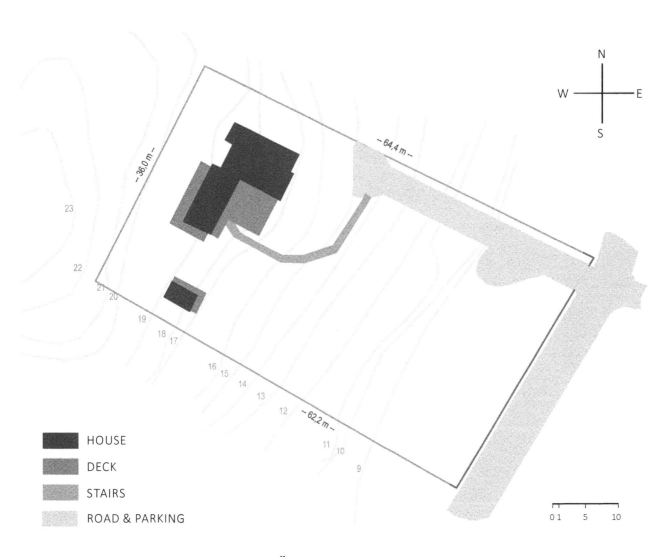

Ängsvägen today

as an additional water source during dry summers. During a typical summer, the water requirements here are not that large due to the natural moisture of the meadow. Lastly, they want a small marsh with reeds in the east corner, to capture as much of the nutrients in the water as possible before it leaves their land.

The meadow has some problems with overflow, especially in the eastern part during periods of abundant rain, and during the spring. This is partly solved with a newly created drainage on the east edge. However, the meadow would still benefit from a regulation of the groundwater from further up the plot.

Currently, there are patios in the form of wooden decks on both sides of the house. The patio outside the kitchen is mostly covered, and there is also the possibility to cook outdoors there. This patio they want to expand somewhat.

The site has no particular wind problems, and the risk of fires is average.

Ängsvägen's weather

Month	Rain	Average temperature	Range	Majority of the winds coming from
January	50	-3	fr -20 to +3	N, NW, NE
February	25	-4	fr -25 to 0	N, NW, NE
March	50	0	fr -25 to +3	N, NW, NE
April	50	4	fr -10 to +15	W, E
May	25	11	fr -2 to +25	SW, SE
June	50	14	fr -3 to +25	S, SW, SE
July	25	17	fr +10 to +30	S, SW, SE
August	50	18	fr +10 to +30	S, SW, SE
September	50	14	fr +5 to +25	W, E
October	75	9	f -1 to +15	W, E, NW, NE
November	75	5	fr -5 to +10	N, NW, NE
December	50	0	fr -15 to +5	N, NW, NE
TOTAL	**575**			

Property facts of Ängsvägen

Altitude	20,9 m to 32,9 m
Latitude	59.2820727
Longitude	18.645366100000047
Distance from the sea	1 000 m
Distance from lake	300 m
Land area	Approximately 2 400 sqm
Access	Through the lowest part of the plot
Building	A converted summer house – built in 1968
Location	Southern location with shade mainly on the north and northeast side of the house
Climate	Temperate with cold winters
Growing season	On average, 180 days
First frost	End of October
Last fros	Early June
Last unusual fros	End of June
Cultivation zone	2–3
Soil	Mainly clay
pH	4–7
Highest temperature	30
Lowest temperature	-27
Wind record	24 m/s
Fire hazard	Minimal
Solar angels	See illustrations
Winter sun-times	07:46 — dawn
	08:42 — sunrise
	11:44 — mid-day
	14:47 — sunset
	15:42 — twilight
Summer sun-times	01:58 — dawn
	03:30 — sunrise
	12:48 — mid-day
	22:06 — sunset
	23:37 — twiligth
Waste and rubbish	90 % is composted and recycled, the rest is collected by municipal authorities
	Septic tank emptied by tanker truk and infiltration bed from greywater
Water	Own well
Extra water	Hand pump in the area
Additional water resource	Adjacent lake
Energy	Electricity grid
Heating	Electricity, fireplace and air heat pump

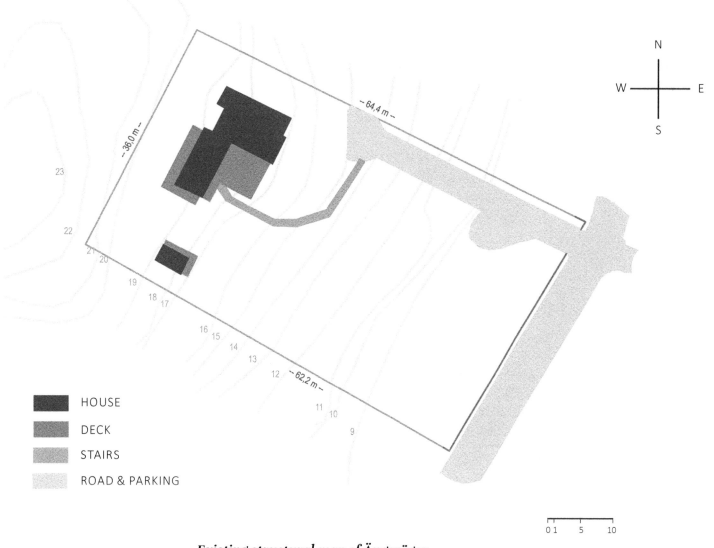

N
W —— E
S

HOUSE

DECK

STAIRS

ROAD & PARKING

0 1   5   10

Existing structural map of Ängsvägen

Site structures

In a design solution, the more or less fixed parts are called structures. These can, for example, be residential buildings and other buildings, balconies, roads, aisles, bridges, fences, and so on. Planning the structures is always about finding one optimal solution that benefits the whole. You do this by deciding which functions you want while taking into account the resources that are required, if they are available, and whether any structure can generate revenue in the long run.

If you already have a plot with existing structures, you need to ask yourself if there are any missing or if there are any that do not work – and therefore need to be altered, or removed. Moving a structure can pay off – if it makes the overall solution better, and if the cost is worth it.

If you start with an undeveloped plot, it is essential to think carefully from the beginning. It is a big project that takes its time because there are many questions to answer. Some examples of questions are: Where should the house be for optimal function? Where should the entrance best be placed for excellent access all year round? Do you need a garage or other buildings, and where should they be placed? What should be grown, and where should the gardens lie? How should the walkways be set to connect all the parts?

The best way is to work out several different solutions in the form of maps that you can then compare with each other. If you already have structures that you want to keep, you first make a basic map with these, before you start sketching on different solutions. It may feel like unnecessary work to create several options. Still, strangely

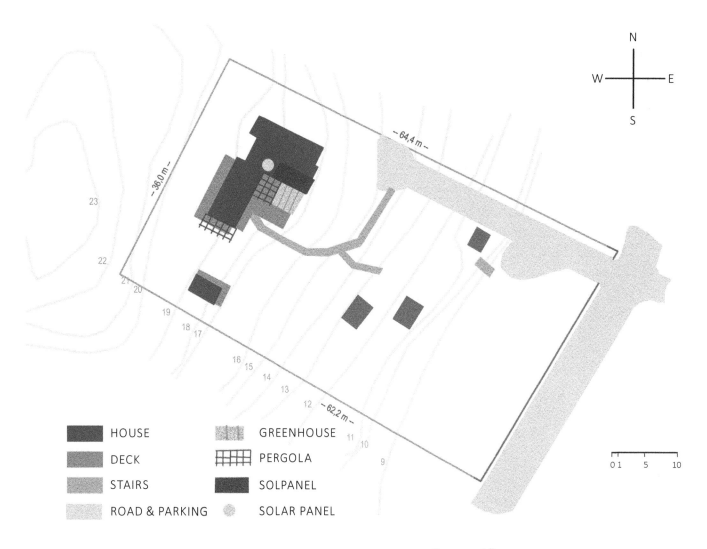

New structural map of Ängsvägen

LEGEND:

■ HOUSE		▦ GREENHOUSE	
■ DECK		▦ PERGOLA	
■ STAIRS		■ SOLPANEL	
■ ROAD & PARKING		● SOLAR PANEL	

enough, it will make it easier to choose and make decisions if you have.

Ängsvägen's structures

The existing structure consists of a single-storey, rebuilt summer house of approximately 100 square meters. The house has a kitchen, living room, newly built bathroom, toilet and four bedrooms – one large and three small ones. A cellar is located under the southeastern part of the house, where one part is used for gardening tools, and the other for general storage. This cellar is ideally suited as a food cellar.

Since the plot is mountainous, there are several stairs up to the house leading up to a wooden deck that extends around almost the entire house.

There is also a small, uninsulated guesthouse south of the house which is currently used as a storage room.

Sewage and rubbish disposal are available. The toilets are connected to a septic tank, and the greywater enters an infiltration bed. Ninety per cent of rubbish is composted or recycled; the rest is handled by municipal waste collection. On the north side of the house, there is a newly built woodshed and a kitchen compost.

New structures

At the house, there is an added greenhouse, two pergolas, solar collectors and a solar chimney on the roof. See more in the detailed description. A new wooden deck between the house and the guest cottage is built to connect the two existing wooden decks.

In the lower part of the forest area, a duck coop and a chicken coop are to be built, both with electric fences around them.

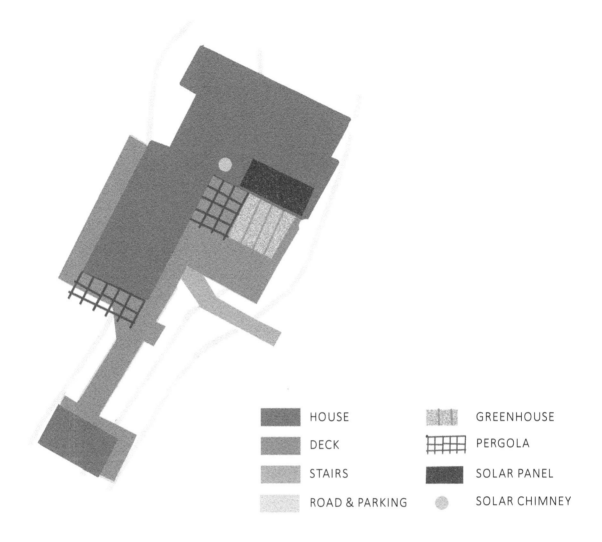

	HOUSE		GREENHOUSE
	DECK		PERGOLA
	STAIRS		SOLAR PANEL
	ROAD & PARKING		SOLAR CHIMNEY

Structural details of Ängsvägen

The house designs are such that the animals can live there all year round. Electricity is drawn from the house because the area is too shaded to use solar energy. Both houses also have extra space for – among other things – feed.

A toolshed is built above the lower parking lot for storing garden tools and the like, to make caring for the crops in the meadow easier.

Additional walkways are created with stairs, both to get a shorter path that goes directly to the animals, and to get an alternative way down to the meadow.

Structure details

The detail plans show the greenhouse, the pergolas, and the solar panel. Existing roofs are raised by about 30 centimetres in connection with the solar panel installation. This extension will provide extra insulation for the home. This increase in the roof height of the house will also make the ceiling height in the greenhouse taller. The solar energy will be able to provide hot water during the sunniest six to eight months.

The greenhouse means an extended growing season while making fruit and vegetables easily accessible from the kitchen. Between the greenhouse and the house, a pergola is created, which together with the greenhouse gives a warming effect in the winter, and a cooling effect in the summer.

On the southern short end of the house, a pergola is built for climbing blackberries and the necessary shading of the wall during hot summer days.

Indoors and to the right of the house, a solar chimney is installed. This is combined with supply air vents on the north side of the house. Both the vents and the solar chimney are designed to be kept closed during the cold seasons.

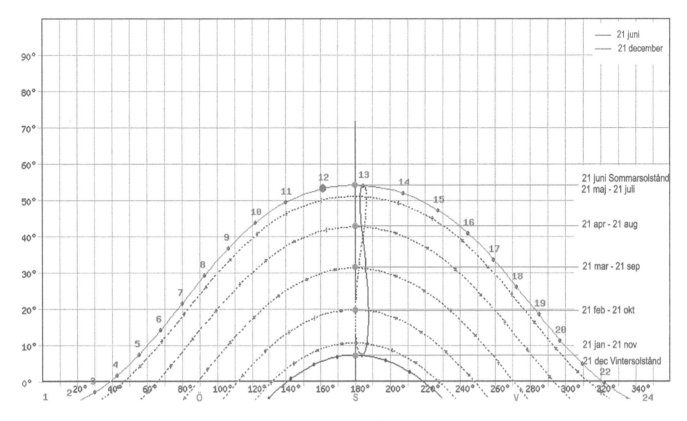

21 juni
21 december

21 juni Sommarsolstånd
21 maj - 21 juli

21 apr - 21 aug

21 mar - 21 sep

21 feb - 21 okt

21 jan - 21 nov

21 dec Vintersolstånd

Ängsvägen's solar altitudes

Sectors to consider

A sector is an area with specific conditions. For example, it can be a place that is sunny, sheltered, or windy. Finding out which different sectors a site has and then marking them on a map makes it easier to create a good overall solution. This section covers the most common sectors.

Sun sectors

Sun sectors are the parts of the plot that have sunlight from morning to evening. Here in Sweden, the size of this sector varies greatly depending on whether it is summer or winter. This is clearly seen in the pictures on the next page, with the summer solstices at the top, and the winter solstices at the bottom. By finding out the boundaries of the sun's movements, and how high the sun is in the sky during the summer and winter, you can see how many hours of sun a place gets. Valuable information in many ways; for plants, animals, and people – as well as from an energy perspective.

Above you see Stockholm's sun angles with winter angles to the left, and summer to the right. If you want to find out your solar times, you can do so on the Sun Earth Tools website.

Wind sectors

Wind sectors are the areas where certain winds have an unusually large impact. If you have lived for a long time in a place, you probably have a good grasp of how the winds move over the plot. If not, you need to observe the winds over at least one year to get an answer to common questions such as; From where do the harsh storms in the autumn and the cold winter winds come? From where do the cooling breezes and the hot, dry summer winds come? Could a specific spot be suitable for a small wind shelter? Could you plant trees and shrubs, or build a trellis to create wind protection in a windy place?

Viewing sectors

Viewing sectors are places where you want to be able to sit and enjoy nature. Let this take some time to figure out, and be happy to go exploring in different seasons. This makes it is easier to discover beautiful places to be preserved. Who knows, maybe there are hidden gems on your plot that could become beautiful viewing sectors.

Noise sectors

In more densely populated areas, there is often "noise" that one wants to avoid. Find out what type of noise it

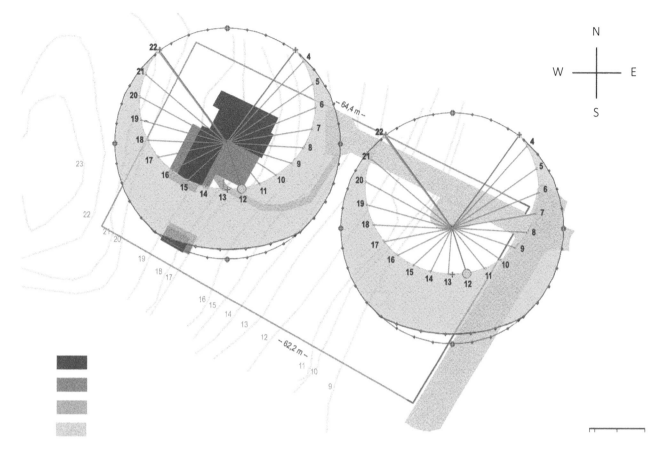

Ängsvägen's solar altitudes in summer

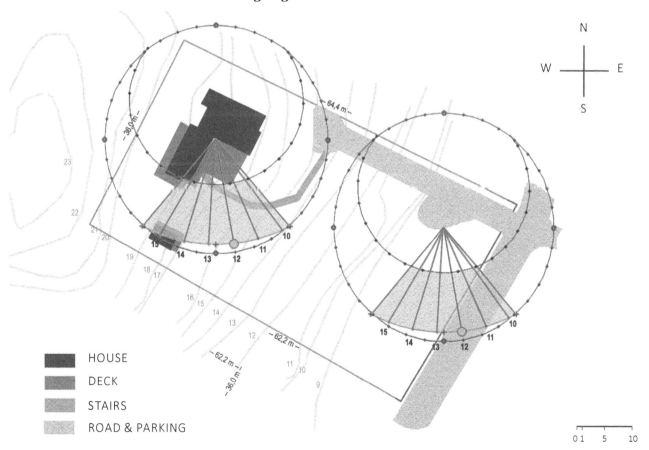

HOUSE
DECK
STAIRS
ROAD & PARKING

Ängsvägen's solar altitudes in winter

is, and what part of the plot it disturbs the most. If the sound needs to be reduced, a fence or the like can be a good solution, but be sure first to check what rules apply to your area. Another solution is to plant dense plants, which are fantastic at stopping the noise.

Dust sectors

Dust and debris is something no one wants in their garden. When knowing which direction the dust comes from, a good solution can be to plant shrubs and trees in that area – because the winds swirl less around them than around a fence.

Odour sectors

In an area, there are always lots of smells to consider. In some areas, there are lovely scents that make us stop and enjoy the moment. In other areas, there are smells that you would rather avoid. They can come from the land itself, or with the wind. Depending on what smells it is,

one can create different ways to both reinforce and reduce the odours.

Flooding sectors

A flood sector is an area where water sometimes overflows. This is usually seen as a problem, but as said before – a problem can be the solution to something else. To find the best solution, it is vital to observe what is happening, and what opportunities are available on the site.

One should, at the same time, remember that the water brings plenty of extra nutrients. One solution may be to plant plants that thrive on being flooded at regular intervals. If there is enough soil moisture all year round, a beautiful and useful solution can be a pond with water-loving plants surrounding it. If it is not possible to solve the problem in place, it should be examined whether the excess water can instead be connected to the rest of the water flow over the site. This way, both the water and its nutrition can be useful elsewhere.

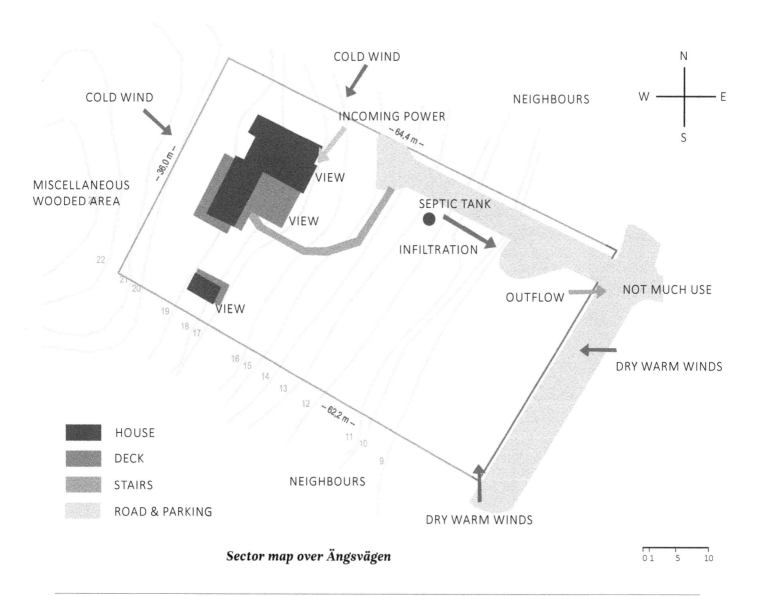

Sector map over Ängsvägen

Gardening in different zones – practical tips

As I explained earlier on, permaculture divides up a plot in several different areas; so-called zones, to optimize energy consumption. The zoning is done based on how often you visit the different parts of the plot. To give a practical idea of what this means, I will go through what is appropriate to have in each zone.

Zone 0

Zone zero is the energy-efficient residential building.

Zone 1

Zone one is the most intense of the outer zones. For a family with two adults and two children, Approximately 1000 sqm is enough for this zone. The zone should not consist of more than 50 per cent hard surfaces and buildings, i.e. garage, roads, barns, and more. Selected parts of this list are also suitable for smaller urban cultivation.

→ *Vegetable Garden*

→ *Things that need harvesting every day*

→ *Things that need to be easily reached for cooking the area is grown for four to six months per year*

→ *Small fruit trees (dwarf trees) and trellis trees*

→ *Mother plants for plant propagation*

→ *Valuable difficult plants*

→ *Trellis for height and shade cultivation*

→ *Deciduous plants*

→ *Place for seedlings*

→ *Quiet animals that are fenced in; for example compost worms, rabbits, quails, guinea pigs, and pigeons*

→ *Water tanks, ponds, rocky outcrops, and herb gardens*

→ *Clotheslines*

→ *Food storage for animals*

→ *Seed storage*

→ *Utilities, workshop, and garage*

→ *Should be fenced in to prevent both one's own and wild animals from entering*

→ *No animals should be allowed to move freely*

→ *On the outer border between zone one and zone two is a good place to have your compost, chickens, and ducks*

Zone 2

→ *Food forest or a forest garden with plants that need attention quite often*

→ *Main crops such as potatoes, pumpkin, grains, sweet potatoes, carrots, and more*

→ *Trees and shrubbery, energy forestry which is regularly harvested for smaller firewood, fence material, and more*

→ *Trees for deciduous harvest for the animals*

→ *Animals can move freely, but only when fruit trees and other plants have grown to the point that the animals cannot harm them. If one is short on land, one alternative may be to fence in each tree. If you want to grow potatoes (for example), you have to choose whether you wish to close in the animals or the crops*

Zone 3

→ *Part of the food forest continues here in the form of trees that take care of themselves.*

→ *Here, too, grass-eating animals are a good idea*

→ *Suitable for growing plants that can be stored. When you have the harvest you need, the animals can eat the rest*

→ *If you want to grow grain or the like, this is a suitable area*

→ *In this area, the forest also begins to enter from zone four*

Zone 4

Zone four is located quite far away from the house, and it is an area you do not visit that often. It is common for zone three and four to overlap.

→ *The most common use of zone four is some form of forestry activity for lumber, firewood, foliage, and more.*

→ *This is a good area for growing plants for pollinators such as bees and bumblebees*

Zone 5

Wild nature, either your own or public – which is rarely visited. If it is your private land, one can, to some extent, strengthen the environment for the wild utility animals to increase their population there. This can be done, for example, by setting up different habitats and houses for birds, butterflies, bats, etcetera. Other than that, it is best to let this part of nature take care of itself.

The zones of Ängsvägen

This site is in many places so steep that it is not appropriate to grow anything in these areas. This means that the location of zone two is less energy efficient than desired, which the owners are well aware of.

Zone one is used daily in the summer. There are seating groups and smaller cultivation areas.

Zone two consists of two parts. Part one is a smaller cultivation area for herbs and some berry bushes with close access to the kitchen. The plot's main cultivation area, part two, is located on the meadow in the south-western part of the plot.

Zone three consists of three parts; the first is the food forest, the other is wood storage and kitchen compost, and the third part consists of the animal areas.

Zone four is kept natural with some large pine trees and some smaller trees with oak, rowan, and juniper, among others.

There is no zone five on the plot. On the other hand, there is a smaller forest northwest of the plot, but it is difficult to reach because of a tall cliff between the plot and the forest.

HOUSE
DECK
STAIRS
ROAD & PARKING
GREENHOUSE
PERGOLA
SOLAR PANEL
SOLAR CHIMNEY
WATER BARRELS
DOWNPIPES
AIR SOURCE HEAT PUMP
GARDEN BEDS
RHODODENRON
VEGETABLES
BERRY BUSHES
CHERRIES

Zone 1 of Ängsvägen

Zone 1

In the newly built greenhouse, tomatoes, chilli, and cucumbers can be grown. Grapes are also suitable. Once established, they will be able to climb the roof and not only provide plenty of tasty grapes but also much-needed shade during hot summer days.

At the pergola between the greenhouse and the entrance of the house, it is good to cultivate a more durable variety of grapes which, in the long run, provide shade.

At the second pergola on the short side of the house, a couple of plants of climbing thorn free blackberries would be ideal. Here, too, the shade is valuable during hot summer days, especially considering that the master bedroom is located beyond this south wall.

The cultivation beds should be elevated according to the sketch, considering that there is a lot of rock near the surface of the soil. To continue to build good soil, mulching is appropriate, provided that a snail barrier can be created.

In this zone, it is suitable to grow what is needed on a daily basis; such as lettuce, sugar peas, spinach, dill, parsley, carrots, radishes, and strawberries.

There is already seating on the wooden deck in front of the house. Depending on how the family wants to organize its cultivation in the greenhouse, seating can also be created inside of it. On the northwest side of the house, there is a shaded seating area for hot days.

Zone 2

Due to the rugged nature of the plot, zone two is divided into two different areas. The smaller part is located on the back of the house where the ground mainly consists of rocky hills. This is a good spot for small cultivation areas for drought-resistant plants – for example, different types of herbs. Despite the harsh environment, there are advantages here – the cliffs radiate heat to create an unusually warm microclimate, and the area is close to the kitchen.

The other part is on the meadow at the bottom of the plot. Here, the ground is too water-retaining, therefore raised beds are also suitable here. Here too, a snail barrier is preferred. Much of the material for the construction of the beds can be found in the wooded part of the plot and in the surroundings. The proposal below is for one four-year crop rotation spread over eight beds – that is, two beds per cultivation group. Between the beds, white clover can be grown, both to attract pollinating insects, and to be used as mulching material in the cultivation beds.

On the other side of the small road that runs along the meadow, raspberries can be grown in double rows. Possibly, a protective hedge can be planted between the raspberries and the road to provide privacy.

Bordering the lower parking area, a more substantial trellis for growing climbing peas and beans is ideal. The trellis also creates a protected spot, perfect for a sitting area.

Suggested crop rotation

Year 1 – Plants that nourish
The first year, legumes are grown, which feed the soil with the help of their nitrogen-fixing bacteria. At the beginning of the season, you can add grass clippings or liquid manure. When you harvest the peas and beans, the haulm is left behind to eventually become humus.

Year 2 – Plants that require a lot of nutrition
In the second year, natural fertilizer is supplied for the cultivation of nutritionally demanding plants such as cabbage, celery, leek, squash, and pumpkin.

Year 3 – Moderate nutrition
The third year, a little natural fertilizer is added for the cultivation of plants that are not as nutritionally demanding. Some can even get worse properties due to too much nutrition. Suitable plants are onions, lettuce, root vegetables (except potatoes), parsley, and dill.

Year 4 – Little nutrition
In the fourth year potatoes or artichokes are grown. No fertilizer needs to be added. However, one can happily plant some leguminous plants – like peas – between the rows.

Zone 3

In the upper part of zone three, it is good to build a small so-called "food forest". Some existing trees need to be removed to let the sunlight in and to leave room for fruit trees and berry bushes. Even part of the plot's beautiful rhododendron shrubs needs to be removed – or at least moved. The lower part of the forest area is left in the existing condition for shade and protection for ducks and chickens. In the border area between the lower forest part and the meadow, there is room for different kinds of berries.

Shrubs and trees that can be grown along the proposed swales are:

→ *Five different kinds of apple trees; a summer apple, an autumn apple, and three different winter apples*

→ *Three different kinds of plum trees*

→ *Two different kinds of peach trees*

→ *Two different kinds of pear trees*

→ *Five American blueberry bushes of at least two different varieties*

→ *Two red currants and two red gooseberry bushes up in the "food forest"*

→ *Four rhubarbs, four currant bushes of different varieties and two different kinds of gooseberries in the area at the bottom of the meadow*

Under the large oak, another sitting area can be created by levelling the ground. The area is both inviting and provides shade. Complementing the area with six rosehip bushes increases the feeling of space. It gives the added bonus of useful and flavourful rosehips.

In the middle part of the forest area both chicken- and duck coops are built, which are fenced in. Beside each animal house, a smaller space is fenced with a higher net

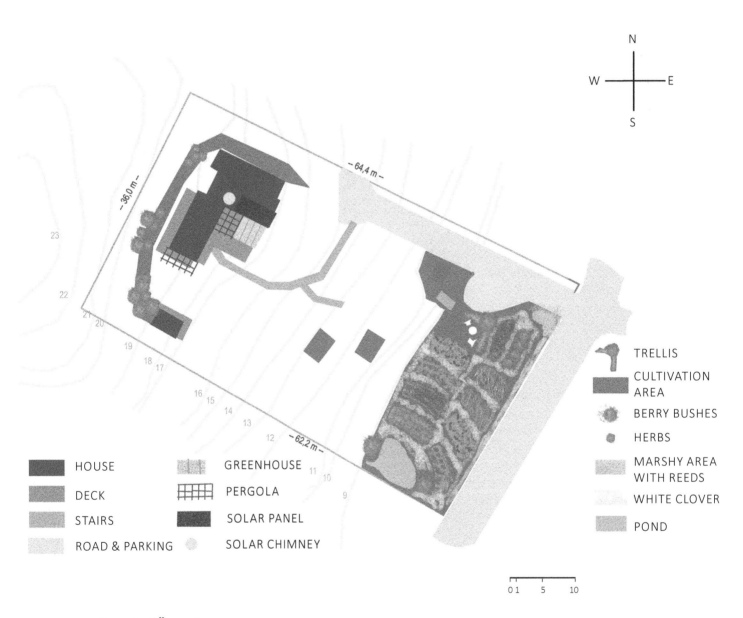

Zone 2 of Ängsvägen

HOUSE
DECK
STAIRS
ROAD & PARKING
GREENHOUSE
PERGOLA
SOLAR PANEL
SOLAR CHIMNEY

TRELLIS
CULTIVATION AREA
BERRY BUSHES
HERBS
MARSHY AREA WITH REEDS
WHITE CLOVER
POND

with a netted roof is made to protect against foxes and birds of prey. With such a solution, the animals can be outside at night during the warm season.

Garden waste that the animals can eat; such as haulm of different kinds, can be more easily carried here from the gardens both above and below the animals by creating a path with simple steps between the different parts of the plot.

Zone 4

Zone four is rarely used due to the fact that the soil is not suitable for cultivation. Along the northwestern boundary of the plot, there are only rock hills.

In the forest areas south and southeast of the small guest house, there is a smaller area with wild blueberries, and along the entire driveway, both wild strawberries and blueberries grow.

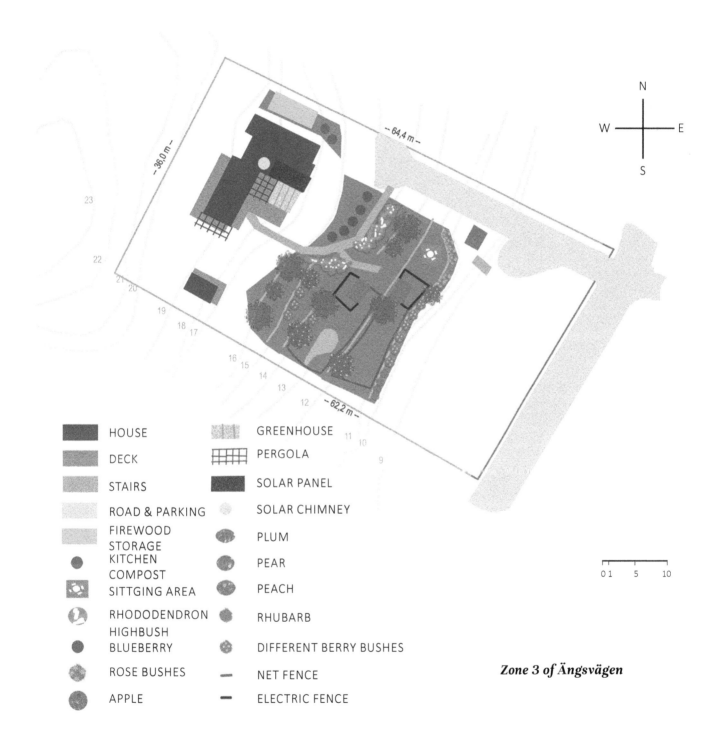

■ HOUSE	▨ GREENHOUSE
■ DECK	▦ PERGOLA
■ STAIRS	■ SOLAR PANEL
▨ ROAD & PARKING	○ SOLAR CHIMNEY
▨ FIREWOOD STORAGE	● PLUM
● KITCHEN COMPOST	● PEAR
▨ SITTGING AREA	● PEACH
◐ RHODODENDRON	● RHUBARB
● HIGHBUSH BLUEBERRY	● DIFFERENT BERRY BUSHES
● ROSE BUSHES	— NET FENCE
● APPLE	— ELECTRIC FENCE

Zone 3 of Ängsvägen

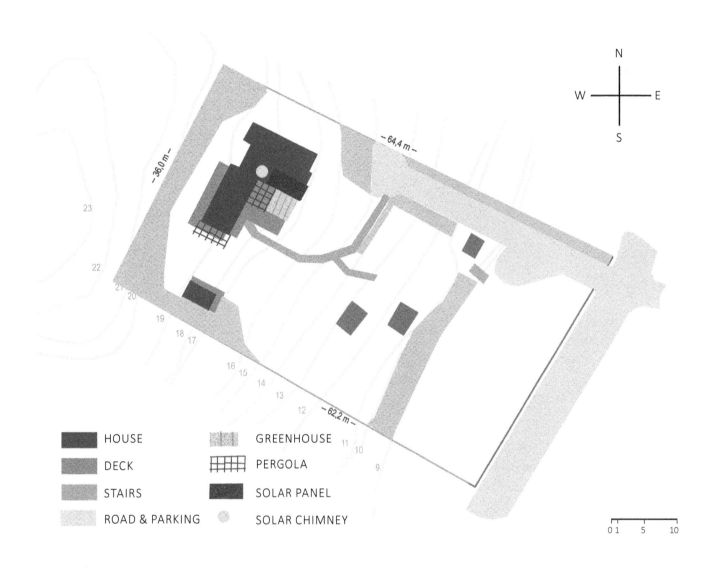

N

W   E

S

– 64,4 m –

– 36,0 m –

– 62,2 m –

	HOUSE		GREENHOUSE
	DECK		PERGOLA
	STAIRS		SOLAR PANEL
	ROAD & PARKING		SOLAR CHIMNEY

0 1   5   10

Zone 4 of Ängsvägen

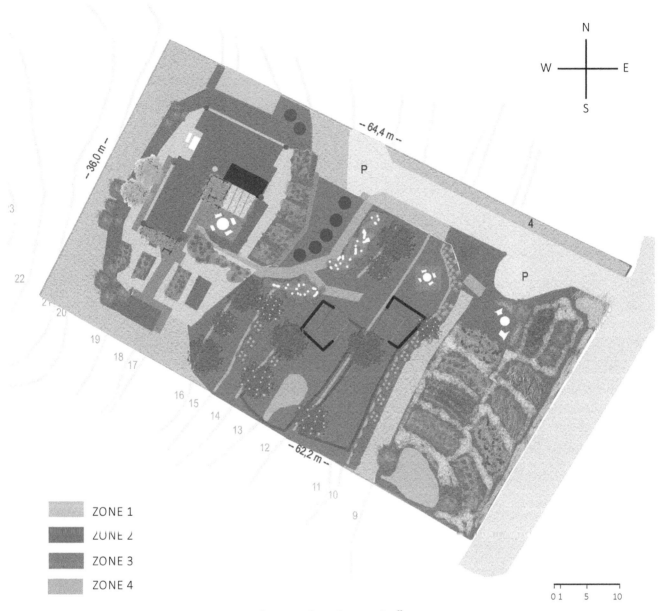

N
W   E
S

P

P

~ 64,4 m ~

~ 36,0 m ~

~ 62,2 m ~

ZONE 1
ZONE 2
ZONE 3
ZONE 4

0 1   5   10

The complete design forÄngsvägen

Practical know-how about elements, function, and contacts

In the chapter "Permaculture" I briefly went through various elements and what functions they have and which contacts can be found. First, a little repetition of this before I proceed with the practical.

Elements

When designing, it is mainly these elements that you work with:

→ *People*

→ *Plants*

→ *Animals*

→ *Water*

→ *Soil*

→ *Structures such as houses, walkways, garages, and roads*

Features

Each element then has several different features that one must take into account:

→ *Needs*

→ *Behaviour*

→ *Products*

→ *Neighbourhood character*

Contacts

Different elements and features have a variety of contact routes:

→ *With what does an element best fit?*

→ *How do the contact routes look?*

→ *How many contact routes are there?*

→ *Is there any contact routes that connect more than two elements?*

Elements and function in practice

A simple and easily understandable example of elements and function in practice is chickens.

The needs of chickens

→ *Fresh air*

→ *The right temperature*

→ *Water and food*

→ *Sun and shadow*

→ *Gravel and shells*

→ *Sand for sand baths to keep fleas away*

→ *Checks that, for example, perches, nest, and enclosures are in good condition*

→ *Protection against attacks*

→ *Chickens need a smaller and more protected area*

→ *Chickens quickly dirty their drinking water; therefore
the water solution must be designed so that there is always fresh and clean water*

→ *Fresh green keeps them healthier*

→ *Health checks and potential treatment*

Products from chickens

→ *Eggs*

→ *Manure*

→ *Meat*

→ *Heat*

→ *Gas*

→ *Feathers (protein-rich, easily composted so that the protein becomes nitrogen)*

The behaviour of chickens

→ *Scratches the ground*

→ *Eats seeds and insects*

→ *Lays eggs*

→ *Roosting*

→ *Clucking and cackling*

→ *Fighting (crowded, or too many roosters)*

The inherent character of the chickens

→ *Breed-specific behaviours depending on the climate; such as feather thickness, leg length, and flight ability*

→ *The size of a chicken determines how many eggs she can incubate*

- → *If a chicken has started to incubate and has remained for 24 hours, she usually remains until the chickens' hatch*

- → *The chicken turns her eggs every day. If you want to save fertile eggs while waiting for a chicken, you have to turn them yourself every day*

- → *A rooster can handle about 10–15 chickens for maximum egg fertility*

Different types of contacts

There are many different kinds of contacts. When it comes to contact between people and other elements, personal energy consumption is an especially important factor. This is easiest taken into account by – before deciding where to place something – calculating how often, and how far you need to go to move between different elements.

Chickens need to be looked after and managed twice a day, which means 730 visits a year. If you then place the chickens 500 meters away from the house, it will mean walking 730 km per year, which is obviously not optimal. As such, it is better to have the chickens a little closer – but at the same time not too close, with regard to flies and noise.

If you instead look at the forest element and how often you visit it, the numbers will be very different. As it takes 25 – 50 years for a large timber tree to grow, the number of visits per year is few. The first few years, trees need a little more supervision – but spread out over the lifetime of the tree; this still only amounts to around five visits a year. With this in mind, it feels obvious that the forest will be farther out in the cultivation system.

Chickens and their surroundings

To give you another example, I return to my favourite – chickens. When you know the function of the chickens, it is time to combine them with the rest of the system. Here are some aspects to consider.

On a sloping plot, chickens are placed above the gardens, if possible. This is done because the nutrients that the poultry farm delivers to the ground follows the groundwater down the slope and thus fertilizes the crops. This is good to think about even in other contexts – because one can always rely on and use the water as nutrition- and energy transport by utilizing earth's gravity.

When on sloping ground, chickens always scratch away at the ground so that anything they move falls downwards. This can be exploited by putting a gate on the lowest side of the enclosure to easily access the compost that over time collects there.

Another way to utilize chickens and their scratching is when arranging for new crops. By having a smaller, removable hen house and a simple removable fence, you can let the chickens process the soil in the intended area. You start by placing the hen house and fence in the area's lowest part. When the chickens have worked it over, all you need to do is move the whole carriage a step forward and upwards. Behind and below, the ground surface is left both processed, fertilized, and ready for planting. This solution is commonly called a "chicken tractor".

Chicken tractor

Another way to use chickens and their natural behaviour is to build a combined chicken and greenhouse. Here, the heat from the hen house can be used to heat the greenhouse. If you want, you can also let the chickens, at least periodically, into the greenhouse. However, this requires some special arrangements with, for example, raised cultivation beds where the chickens have their perches below the beds. Partly because heat rises, but also to prevent the chickens from scratching around in your beds. See more in the chapter "Extending the season".

Trees, pastures, and animals

Combining a pasture with trees and shrubs increases and improves yield from the area. Cows love the scent of fallen fruit! When the cows walk around and eat their apples, the soil is processed, while the trees are fertilized. The trees also provide animals with protection against both sun and bad weather.

The nutritional competition that arises between trees and grass is counterbalanced by the trees shedding their leaves in the autumn, thus providing a supplement of nutrition.

When the grass is not enough, some deciduous trees, such as ash, can produce leaf fodder for the animals. This is done by severely cutting (pruning) the tree and leaving the leaf-clad branches lying on the ground. Another benefit of the solution is that you do not have to transport your extra feed.

Shrubs that are correctly placed also creates protection for animals and plants. Rosehip bushes are another excellent example of a plant with many functions. They are beautiful, produces antioxidant-rich rosehips and – thanks to their thorny branches – are an environment where many wild animals feel comfortable, such as birds and insects. These animals, in turn, benefit the pollination so that the harvest of fruit and berries is secured.

Your imagination sets the limits

There are an outstanding amount of ways to combine different elements and their various functions. It is practically only your imagination that puts an end to it. For problems, it is often just a matter of perspective, where a problem can prove to be the solution to something else! Again, I would suggest making contact lists, where you use a line to connect the different parts as seen in the picture below.

Looking for empty spaces in a system, something I talked about in the chapter "How permaculture works", comes in handy here. Through careful observation and notes, one learns to see what different elements, functions, and contacts are missing. To repeat this below is the text about the different spaces again.

Different spaces

There are three types of spaces to look out for:

1. *A space in places, which can be a place to be in, to fit in, and find food in – a place with a roof over your head, space to function in*

2. *A space in time that may be cycles of recurring occasions – such as annual, seasonal, monthly, weekly, and daily*

3. *A combination of both of the above, that exists when time and space coincides*

A few examples are:

→ *Vertical spaces*

→ *Location*

→ *Different Zones*

→ *Different soil types*

→ *Different depths of water*

→ *Different slopes*

→ *Different forms of flow through a system*

→ *Boundaries between edges and margins*

Co-cultivating with trees for a sustainable garden

Our trees play a vital role in life here on Earth. Thus they are essential components, both when gardening and keeping animals. This chapter begins with a summary of the trees' functions. At the end of it, I will go through things to consider when buying and planting trees – especially fruit trees.

Functions of the tree

For a farmer and gardener, the main focus is usually the harvest, even when it comes to trees. Often the role of the other trees is forgotten – for trees have many more functions than just providing a harvest. By including trees that have several different functions in your planning, the harvests will not only yield more, but the whole ecosystem will thank you.

Trees collect water

As I explained in the chapter "Without trees, we would have deserts", the trees play a vital role in nature's water cycle. Here is a brief summary:

→ *Rain along the coast contains 100 per cent seawater. The farther you get from seas and lakes, the more of the water in the rain is made up of water from the trees. Far away from larger bodies of water, there would be no rain without trees.*

→ *Unlike vapour from seas and oceans, vapour from forests contains considerably more rain creating particles.*

→ *A tree recycles over 74 per cent of the rain. The rest continues down to the groundwater.*

Trees provide shade

The crown of a tree provides shade and coolness for both people, animals, and houses. Their shade also makes it possible to grow a wider variety of plants, as many plants thrive both in partial shade and in full shade.

Trees as shelter

Trees can change the climate of a place. Just think about the difference between a forest and a felled area. By having trees placed thoughtfully, you create adequate wind protection for the benefit of plants, animals, and humans.

Compost from trees

When the leaves fall to the ground, the next step in the ecosystem's nutritional turnover starts. Do not rake the leaves – let them remain and decompose to become food for plants. They then provide the benefit they were meant to. The leaves also effectively help prevent evaporation from the soil. Falling leaves are one of nature's best mulching materials. If you have an excess of leaves in your garden, you can rake some and cover the ground somewhere else, put them in the compost, or give away the leaves to someone who does not have enough. Leaves are full of nutrition and should not be burned.

The leftover branches from pruning are excellent for your compost – and if you cut the branches into small pieces, they break down faster. If there are too many branches, they can be saved in a pile for later use – for new growing beds, for example. In the meantime, hedgehogs like the environment they provide. Larger branches can also be used for firewood.

Co-cultivation with trees creates diversity

The plants that grow beneath and around a tree can – by using a well thought out design – create an interaction that benefits the group of plants as a whole. The shadow of a tree is an excellent example of this; because it helps the nitrogen-fixing plant to thrive, which favours another species that is good at attracting insects, who then, in turn, pollinate the tree.

Trees improve soil quality

The trees also help to reduce and counteract soil erosion by "holding on" to the soil. This works best in natural forests, but worse in planted monocultures where the ecosystem is out of alignment, even if the environment looks like a forest.

Many animals live in trees

The trees offer a diverse habitat for many different animals. Insects are attracted to flowers and fruit; birds are attracted by both fruit and the availability of insects. The trees also give many animals homes with protection against both weather, wind and predators.

Combining fruit trees with different kinds of pets results in even more benefits. The animals that enjoy eating the fallen fruit help reduce pests, due to the fallen fruit being a perfect place for insects to lay their eggs in. Larger animals such as sheep, goats, and cows contribute with nutrient-rich manure while processing the top layer of soil. In turn, the trees protect the animals against the hot sun, harsh wind, and rain.

Trees can reduce energy consumption

Strategically placed trees around the house can save you a lot of energy. This applies in particular to deciduous trees that provide shaded coolness in the summer, and in the winter lets the sunshine through to provide extra warmth in the house. Evergreen trees contribute both shade and wind protection, but these should only grow on the north and north-east side of the house, so that the weak winter sun is not blocked.

Trees for noise protection

Trees are also excellent for noise protection because they sever the sound waves and absorb the sound. Best of all is broad-leaved evergreen plants, but the effects must be weighed against any negative effect these have on the winter sun.

The beauty of trees

Last but certainly not least, it is the trees that largely create the beauty of the landscape. Not only through their statuesque forms; but through the whole ecosystem that becomes possible thanks to these trees. In Japan, the forest has even become a recognized therapeutic concept. It is recommended that you take a so-called "forest bath" to reduce stress and improve mood. Forest bathing means that you take a quiet walk in the forest, or just stay among the trees in peace and quiet. Hanging out with trees has great benefits for us humans.

Buying and planting

The following is advice and tips for buying and planting trees – with a focus on fruit trees.

Best time to buy trees

The best time to plant depends on the type of tree you buy. Those with root balls are best to plant in late autumn or early winter because the tree is then at rest. Trees planted late in the year will also start growing earlier the coming spring. You can also plant in early spring, but the tree will start growing later. Pot-grown trees can be planted at any time during the year.

Prepare the soil

Prepare for the actual planting by dealing with the hole for the plant before purchase. Once the tree is at home, it is crucial that it quickly gets into the soil, due to the sensitivity of the roots.

Sun is the best setting to get an abundant harvest from a fruit tree. The best soil for fruit trees consists of 40 per cent compost soil, 30 per cent clay soil, and 30 per cent coarse sand.

The soil in new residential areas often consists of blasted stone and fillings with a thin layer of topsoil for the grass. This soil is far too poor for fruit trees. Therefore, one must usually dig out quite a lot of this soil, and replenish with a more fitting soil.

If you have clay soils, you mix it with coarse sand, gravel and compost soil, and if it is sandy soil, you instead mix in clay soil and a lot of compost material. Of the purchasable soils, organic soil has the best mixtures.

Regardless of the nature of the soil, it is good to loosen the bottom of the planting pit before continuing.

If the ground is marshy, avoid planting fruit trees there. You can check this yourself by digging a pit and seeing if the water remains in the hole. If so, choose another location. Wet ground both delays the growth in spring and increases the risk of diseases.

If the soil is just a smidgen too wet, or if it is a clay soil, it can be counteracted by setting up drainage at the bottom of the pit. Try and fill up the hole with water; if it takes longer than a few hours for the water to disappear, it must be drained before the tree can be planted. Dig a pit that is at least twice as deep as the root height, and at least four times as wide as the root width. The pit can advantageously be square – or at least not round. This helps the roots navigate more quickly out of the pit, and you counteract root spin – where the roots only grow around and around in the pit. The larger the hole (especially the width), the easier it will be for the tree's roots to find enough nutrients during the first few years. It is preferable to drain the pit with both stone and gravel.

Buying fruit trees

It is worth the effort to spend some time to look over the tree you intend to buy. The site of the inoculation, that is, where the rootstock and fruit trees are grafted together, should be well healed. Stem and crown branches must be undamaged. To be sure that the tree has a strong life force, it should be 60–100 centimetres high and have at least three crown limbs that are well spaced laterally around the trunk at different heights. If the tree has many branches, then make sure that there are three to four good branches to retain after the first pruning. These branch attachments should be strong and should have an angle to the trunk as close to 45 degrees as possible.

A typical apple tree needs a space of about 10 meters in diameter to avoid it growing into anything else when it has matured. If you want a smaller tree, some varieties are smaller when mature. Since it is the tree's rootstock that mainly determines the tree's mature size, it is wise to ask the nursery if the tree you are planning to buy has a weak or strong growing rootstock.

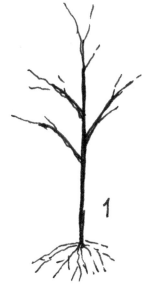

Here's how it should look from the side with the three crown branches at different heights and a straight head.

This is how it looks from the top with three crown branches in different directions.

The branch to the right has a branch angle that is too sharp, which increases the risk of damage when the tree becomes bigger.
The branch to the left has a blunt angle that risks giving fruit too early and thus damaging the tree.

This tree cannot be pruned to form a good crown. Do not buy such a tree.

Planting trees

Prepare the tree by placing it in a tub of water overnight. It will then absorb plenty of water and will not run the risk of drying out. Remove damaged and rotten roots.

An old well-proven planting method for both fruit trees and berry bushes is to plant them somewhat elevated. Start by filling the pit with soil until about 10 centimetres remain to ground level. Tread lightly so that the soil does not collapse when the tree is set down.

A tree needs support at the start, and these supports should be put down before the tree is planted to avoid damage to the root system when the tree moves in the wind. Depending on how windy the location is and how big the tree is, up to three supports are needed in the first year. If the area is not that exposed to wind and the tree is not very large, one support will suffice. This is placed on the side of the tree from which you have the most wind. Place the supports well away from the tree so that branches can grow and move freely without being damaged.

The support can only be there for a short amount of time. If the support is left for too long, the trunk only thickens above the support bracket itself. When the support is removed, the tree risks breaking off at the narrower part. To find out if the supports can be removed, one can press lightly sideways on the trunk. If the soil above the roots does not lift, it is time to remove the supports.

When the pit and supports are in place, it is time to put down the tree. If it is a bare rooted tree, be sure to spread the roots properly. Fill with soil and step on the soil lightly to compress it so that the plant stands firm. The soil around the trunk should be at the same level as when the tree was in the pot, in the sack, or at the first planting site. When finished, the tree should be raised on a small hill, preferably 10–15 centimetres above the original ground level.

Tying up the tree against the support is done with some form of a soft band where the material against the trunk should be about five centimetres wide so as not to cut into the bark. The band should sit just below half the height of the trunk.

Watering trees

How much watering you need to do after planting depends on what type of soil you have. Muddy soil retains water better and requires less watering, while sandy soil needs to be watered considerably more often. Always start by watering the pit by filling it with water. Then wait until all the water has sunk before the tree is planted.

When it comes to a bare rooted tree, it needs more irrigation than one that has been in the soil before it was bought. Namely, the water helps to make the soil particles come into close contact with the fine root hairs.

One recommendation is to give newly planted trees 15 to 30 millimetres per watering, i.e. 15 to 30 litres per square meter. Use the water sprinkler, but make sure that the area it covers is no larger than what is needed, in order not to waste water. Then water slowly and keep track of the amount of water with some form of water meter standing under the spray.

The first few weeks, the tree is especially sensitive to dehydration. So be generous with the water during this period if you are not getting enough rain. Grass clippings work very well as a vapour barrier – but remember not to put the clippings right next to the trunk, as the risk of disease increases.

Establishing time

It can take up to four years before the roots of a new plant have established themselves on the site. An essential part of the adaptation is that the roots must grow sufficiently large and reach deep enough to cope with the winter frost in the best possible way. A good and easy way to increase the plant's chances of survival is to cover the soil for the winter, for example, with straw. This is especially important for the plants that are planted in a plant zone that borders on what they usually can withstand.

Pruning of newly planted trees

Wait to prune a new tree until the tree has been in the ground for a full growing season at that location. The first growing season it needs its leaves to provide energy to establish the roots. If you plant in autumn, the tree is, therefore, left untouched the following year, and is pruned after 1.5 years has passed since planting.

Nutrition for newly planted trees

The best way to nourish a tree is to put down compost in line with the crown's outer circle, as this is where the feeder roots grow.

For the first few years, the tree needs more nitrogen, so it is good to grow different nitrogen-fixing plants such as clover, beans, and peas around the tree. When the summer has come to an end, all of this is cut down and left on the ground. The nitrogen is then released, and the tree is fertilized. The larger the fruit tree, the less nitrogen-fixing plants you need.

Aggressive pruning and trimming

On older fruit trees, a more aggressive trim may be needed. But be sure never to remove more than one third per year; otherwise, you run the risk of killing the tree. This is due to the fact that it is not only the branches that disappear, but the corresponding part of the root system also dies.

One can, therefore, not trim the whole crown of a fruit

tree. To trim is an old way to harvest leaves from deciduous trees as extra fodder for the animals.

Conversely, if you dig out more substantial volumes of the soil under a tree, you then also remove a lot of the tree's root system, which in turn causes the tree to die.

Remember that the tree above the ground surface is only half the tree!

To save and conserve water

It is becoming increasingly evident that we can no longer take freshwater for granted. We must in all conceivable ways reduce our water consumption – otherwise, we risk being left without clean water; both to drink and to water our crops with when the rain is scarce. We also need to learn how to take advantage of rain and groundwater in our gardens and use it optimally.

Control over the water

There are many ways to handle water and to find the best solution in a specific place; you need to look at the whole. Having control over the water flow is important both when it comes to having enough water for prolonged droughts, and to handling extreme weather with large amounts of water.

Saving water

Utilizing different elements and their functions for a sustainable water system is within permaculture, a central design issue that constantly reappears from different perspectives. Perhaps this sounds repetitive, but the water issue is exceedingly essential. Here are some general examples.

The soils ability to retain water

A soil rich in organic matter balances both muddy and sandy soils so that the water passes at a reasonable rate, thus becoming neither too dry nor too wet. Healthy, thriving soil can save you up to 30 per cent thanks to its optimal water retention capacity. Also, a porous soil makes it easier for the roots of the plants to branch out and absorb the water.

Mulching reduces evaporation

Mulching is helpful in terms of water consumption, as the cover material reduces surface runoff, reduces evaporation, and prevents wind erosion. At the same time, the nutrients are released more slowly and evenly.

Local varieties can reduce the need for water

Using local varieties of plants can also reduce water consumption as these plants are used to the local precipitation amounts, as well as to all the other climate effects.

Dew, fog, and mist also provide water

In prolonged droughts and during watering bans, it may be useful to know a few lesser-known methods taken from hotter climates where you are used to using all forms of water, including dew and fog.

One way, as seen below, is to collect stones and stack them sparsely, under a fruit tree (for example). Dew and fog will then condense against these stones.

The best dew collector is a free-standing bush that is 1–2 meters tall. Closely planted shrubs are comparatively bad at collecting dew. If you want more bushes to collect water, they must be sparsely placed so that the fog can slip between them and get maximum contact with each bush. Thus, something to consider when planting berry bushes.

Best time to water

Watering when it yields the best results is a great way to save water. The best watering time is early in the morning until 9 am. Then the plants have time to absorb the water before the evaporation effect of the sun kicks in. Some plants can also cope with water after 4 pm, but keep in mind that the stronger and denser the foliage, the higher the need for the leaves to dry before evening, to minimize the risk of fungal diseases.

Condensation collection

The importance of water temperature

When irrigation water is too cold, the root activity and the plant's capacity to absorb water is inhibited. Therefore, tap and well water are usually too cold. It is better to pour the water into a slightly larger vessel so that it can sit and reach the same temperature as the air.

Avoid overwatering

In permacultural crops, it is easy to overwater before getting used to lower water demands. Although the surface looks dry, there may well be enough moisture a bit further down in the soil. A general rule is that the top two centimetres should be dry, and the soil below moist. This contributes to the roots seeking deeper into the soil, which in turn makes the plant tolerate both wind and drought better. On the other hand, excessively wet soil cannot hold enough oxygen for the plants to thrive, and with too much water, the plants suffocate. Commonly, it is said that 25 to 30 mm once a week is sufficient for most plants. Still, with different water-saving cultivation methods, this can easily become too much. The best way is to check the soil before watering. The most important thing is to make sure that moisture truly penetrates the soil to stimulate the roots' growth downwards. A good thing to keep in mind is that one millimetre of the precipitation corresponds to one litre of water per square meter.

Reusing greywater

Reusing your greywater is a major step, but if you are building or renovating a house, much is gained by thinking about a solution where you separate the greywater from the pure wastewater, or perhaps even choosing a waste-free toilet system. The greywater can then be led to a natural purification bed before it is used for your garden beds. Needless to say, such a system requires that you do not use a lot of chemicals in your home.

Irrigation system

There are many good ways to reduce water consumption when cultivating. Here are some examples.

Drip irrigation

The most effective way to water is to install some kind of drip irrigation. There are two common solutions; one is a long hose that has tiny drip tubes evenly distributed along the hose. These short tubes are inserted into the soil around the plant you want to water. The second variant is a hose which is perforated with numerous small holes. This type of drip irrigation can be placed above ground, but can also be buried in the cultivation beds, as long as the soil is not too sandy. None of these irrigation systems can withstand the frost and need to be protected in some way. Easiest is to clean it properly, roll it together, and put the hose somewhere indoors until the next season.

DRIP LINE

Drip irrigation hose

Self-watering vessels

A 1.5-litre pet bottle can be an excellent self-watering vessel for single sprouts. Divide the bottle in the middle and fill the bottom with some water. The upper part is turned upside down, a wick is inserted, and then this is filled with soil. You can then plant your shoot.

Self-watering pots are available to buy ready-made, but you can easily create one yourself in this way for cheap. How long a plant can survive on one watering depends on the plant and the size of the pot, as well as where it is placed. For example, a tomato plant in a reasonably large pot can survive for a few days, even if it is warm and bright out.

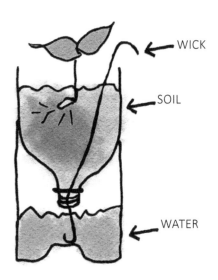

WICK

SOIL

WATER

Self-watering bottel

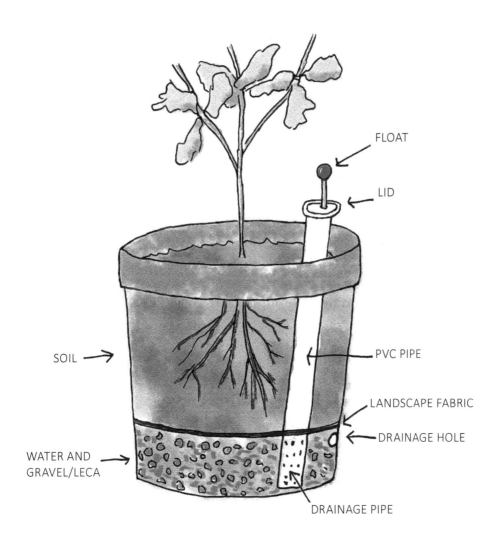

FLOAT

LID

PVC PIPE

LANDSCAPE FABRIC

DRAINAGE HOLE

DRAINAGE PIPE

SOIL

WATER AND GRAVEL/LECA

Self-watering planter

Self-watering beds

If you grow in beds, you can with some additional effort, make them self-watering. Here it is important to get the bed watertight so that the water does not leak out. The easiest way to solve this is to cover the inside of the bed with a pond liner, big enough to reach up to where the soil level starts.

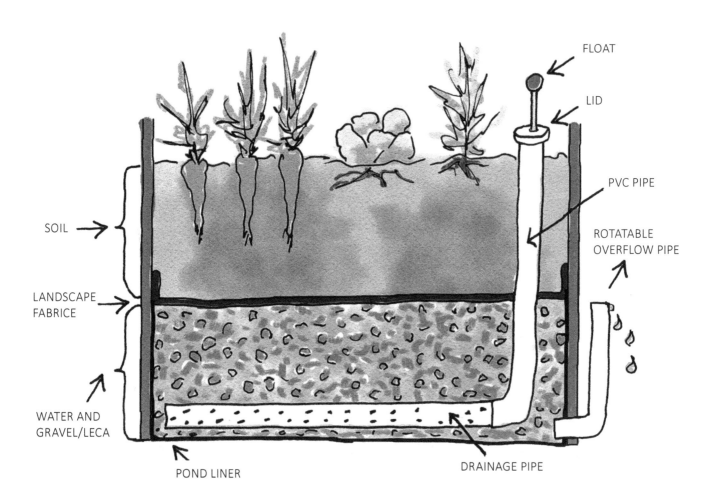

Self-watering bed

Irrigation with drainage pipes

If you need a more extensive system, you can create entire irrigation systems with the help of ordinary drainage pipes. The water to the system can come from several sources; such as rainwater from roofs.

The picture to the right shows how you can easily connect a downspout to the drainage pipe that makes for the start of the irrigation system. To prevent issues with water freezing when the frost hits, the downspout must be protected with a hatch for opening and closing it off. With this simple solution you can, while waiting for the ground to thaw, collect the melt-water from the roof – for example, by placing a barrel next to the downspout.

By digging ditches in the pathways around the cultivation beds, you can create an efficient irrigation system. The image below shows a cross-section of such an irrigation system, where you see the design of the drainage ditches around the beds. The ditches are designed so that they also function as walkways between the beds. The beds in the picture are raised hügelbeds.

Connection between downpipe and drainage pipe

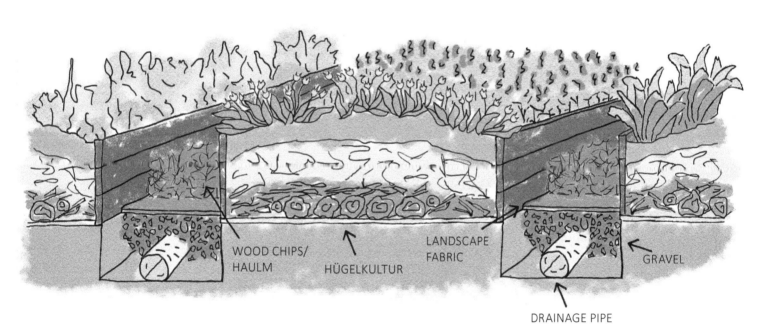

WOOD CHIPS/ HAULM

HÜGELKULTUR

LANDSCAPE FABRIC

GRAVEL

DRAINAGE PIPE

Hügelkultur with irrigation/drains

Drainage irrigation on flat ground

On flat ground, the drainage ditch is dug as in the picture. For the water not to flow through the system too quickly, the ditches should be horizontal, that is, at the same depth. It is good to use some kind of levelling instrument.

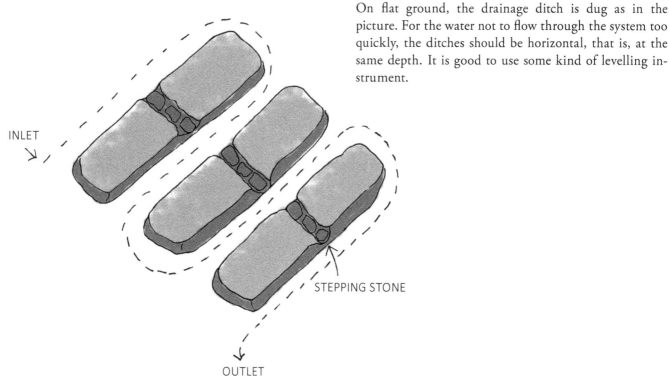

INLET

STEPPING STONE

OUTLET

Drainage irrigation on sloping ground

When the beds are at different levels, the trenches are dug around each bed horizontally. In the transitions between the different levels, a soakaway is made where the water from the upper bed passes through and then goes into the next horizontal drainage ditch.

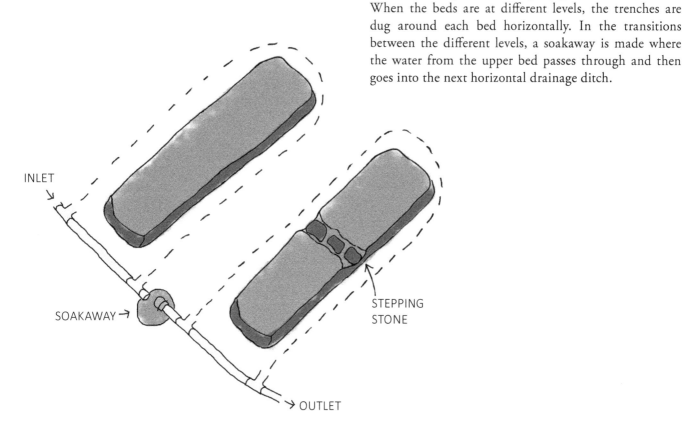

INLET

SOAKAWAY

STEPPING STONE

OUTLET

Soakaways

To prevent debris in an irrigation system, a soakaway may be needed to receive either groundwater, water from a purification pond, or water from an irrigation system at a higher level. Do the following:

→ *Dig a pit of about 80 x 80 centimetres from the upper level to the lower level to create a relatively flat base.*

→ *Cover the hole with geotextile fabric, with enough fabric to later fold over the finished soakaway.*

→ *Insert the drain end of the drain pipe and secure the pipe with a U-shaped nail.*

→ *Cover the pipe opening with geotextile fabric and attach both fabric and pipe with the same type of nail.*

→ *Fill the pit with coarse gravel.*

→ *Fold over the rest of the fabric so that the pit is prop erly covered. Then fold the fabric from below over this.*

→ *Finish by placing some stones on top, holding everything in place.*

→ *The soakaway is then concealed with something rela tively easy to move. It can be a piece of decking, a bench to sit on, or why not some large stones that add to the design aesthetically.*

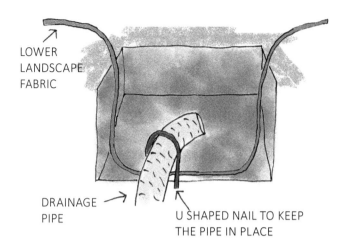

LOWER LANDSCAPE FABRIC

DRAINAGE PIPE

U SHAPED NAIL TO KEEP THE PIPE IN PLACE

LANDSCAPE FABRIC

U SHAPED NAIL TO KEEP THE PIPE IN PLACE

GRAVEL

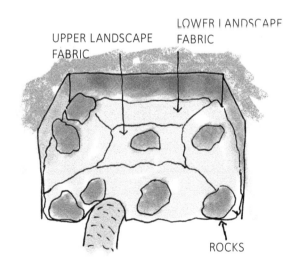

UPPER LANDSCAPE FABRIC

LOWER LANDSCAPE FABRIC

ROCKS

Purification Ponds

Purification ponds may be needed if the water you have isn't good enough for irrigation. It may be rainwater from roofs that you are not sure are clean enough or natural water that needs to be filtered before coming to use. With a slightly larger and more advanced system, you can also purify Graywater water from the kitchen and shower.

If you are unsure of how to proceed, many companies are happy to provide advice.

A dug purification pond on a slope

Digging a smaller purification pond can be a good solution for smaller areas. Here is an example of a pond located on a slope where incoming water enters via a PVC pipe. After the water has passed through the pond, the outgoing water may end up in another small pond, or directly in an irrigation system. Do the following:

→ *Dig a pit of about one square meter with a depth of 50–100 centimetres.*

→ *Build edges at a suitable height.*

→ *Cover the hole with a pond liner so no water can leak out.*

→ *Install a PVC pipe through the liner to drain the purified water.*

→ *Insert filling and drainage pipes, as shown.*

→ *Fill with natural gravel or LECA so that the water depth is about 20–30 cm.*

→ *Plant your plants and top off with water.*

A natural purification pond

If you have a larger plot, an alternative is to build a natural pond that looks something like the picture on the left. The best shape is where the width is at least two meters and the length six meters with a depth of 20–30 centimetres. The form of the recesses in the pond must look like the picture; where the edge of the incoming side is elongated, providing the possibility for underwater plants to thrive. The outgoing side should be steep so that sediment is easily deposited there.

The bottom should consist of sand or natural gravel. You could potentially cover the bottom of the entire pond with a pond liner if you know that the incoming water contains substances that are not good for the environment. For a good result, one-third of the pond surface should be covered in plants.

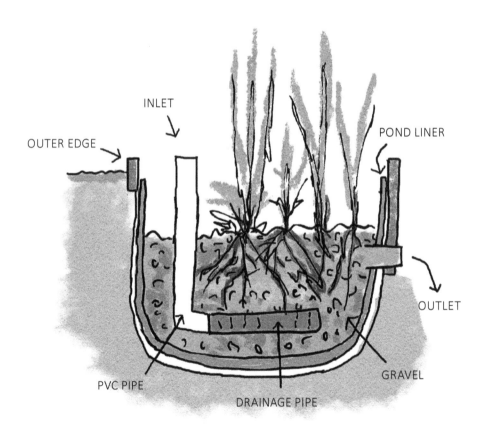

An excavated pond in a slope

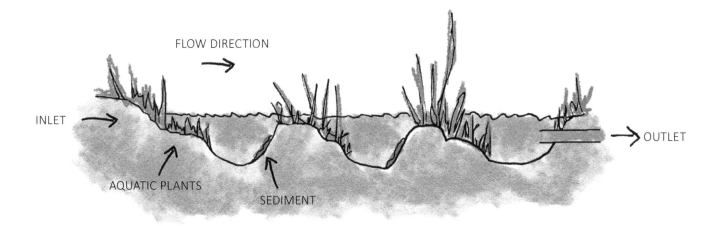

FLOW DIRECTION

INLET

OUTLET

AQUATIC PLANTS

SEDIMENT

A natural pond for water purification

Collecting rainwater

Collecting rainwater is becoming increasingly important as access to groundwater becomes worse. The two main advantages of using rainwater when growing are that you save on drinking water and that you avoid any chemicals added to municipal water. If you live in an urban environment, the city's stormwater system is also spared a part of the rainwater.

Green roofs collect more water

A sight that is becoming more and more common is the planted "green" roofs of old. Here the plants have direct access to the water before it flows down to other collection systems. A green roof is also an effective insulation method, and therefore also reduces heating costs.

Make use of eavesdrip

To collect eavesdrip along a house wall – and at the same time protect the wall against unnecessary water – you can plant a cultivation ditch as in the to the right.

In barrels and tubs

The most common way to collect rainwater is to place barrels under the pipe. The disadvantage of the barrels is that they overflow when there is a lot of rain all at once. This loss can be counteracted by making a drain from the first barrel and connecting this to a lower, and preferably wider vessel. Of course, this can be done in more than one step. If you have space, you can put the first barrel on something so that it sits at a higher level. Another simplifying solution for easier water retrieval is to put a tap on the bottom of the barrels.

INTAKE PIPE IN PVC WITH A LID AND FLOAT

EDGE

LANDSCAPE FABRIC

POND LINER

GRAVEL

DRAINAGE PIPE

Collecting rainwater from roofs

Multiple rain barrels connected

Water tanks

If you live somewhere where there is a lack of natural water, it may be wise to invest in one or several larger tanks where rainwater collects. In larger systems, pumps may be needed. The energy supply to the pumps can be solved in several ways, from everything from household electricity to hand pumps.

To create decent self-pressure, tanks above ground should be located higher in relation to the places where the water will be used. The construction they are put on must be solid to withstand both its weight and all kinds of weather. However, systems above ground must be emptied during winter to avoid frost damage if you live in colder climates.

Underground water tanks are more expensive to install but also easier to collect water to since gravity can handle part of it. Nor is the same scaffolding or wind protection required. A summer system doesn't have to sit that deep, provided it is emptied during the winter. A frost-proof system that is sat deeper down is a significant investment, but then has the advantage of being of use all year round.

No matter what kind of tank you choose, the tank needs to be completely sealed, except for the inlet and outlet. This is to avoid falling debris and mosquitoes laying their eggs in the water. If you choose a tank above ground, keep in mind not to get translucent one, as this promotes algae growth. If the tank lets in light, you need to create shade by either building a small shed around it, or a trellis for climbing plants – or why not a combination of wooden structures, evergreens, and climbing plants.

Rainwater harvesting from roof

Choosing a tank

When choosing the size of the tank or tanks, you need to take into consideration both how much water is reasonable to collect, and how much water may be required during a dry period. If there is an excess of water, the water consumption determines the size.

Choosing materials for different types of vessels is both a matter of size, durability, and cost. On a small scale, this means various forms of barrels of both plastic and wood. But for tanks on a slightly larger size, there are several other materials to choose from, each with its different pros and cons.

Concrete tanks

Concrete is best suited for underground tanks. The disadvantage is that lime from the concrete can leak into the water, especially in the beginning, which is not great for the plants, as most plants prefer slightly acidic water. To avoid this, you can line the inside of the concrete tank. Before deciding, you should check with the municipality if there are any local rules for underground tanks.

Metal tanks

Metal tanks have always been the traditional choice for above ground. They are cheaper than concrete and easier to install. For irrigation, galvanized metal works even if it tends to rust, especially if combined with fittings and tubes of copper or brass. Rust in the tank is not in itself a problem for the plants, but rather an issue for the tank itself. A galvanized metal tank does not last as long. An aluminium tank is significantly better – they are more expensive to buy but lasts much longer.

Fibreglass tanks

Fibreglass is the best material for a water tank, but it is also quite expensive. Everything is a matter of need, size, and price.

Polyethene tanks

Polyethene is the most common material for water tanks today, both large and small. The material has the advantage that it is cheap and easy to transport and install. However, the durability is lowered if the tank is exposed to strong sunlight for a more extended time. One of the most common water tanks for private use here in Sweden is one of 1000 litres in a steel cage.

Collecting groundwater

Collecting water can be done in many different ways depending on the soil conditions at the site. It can be anything from collecting it in ponds to building ditches that meander over the site to decrease the speed of the water. Why not opt for a pond with fish, for example? The more ways you can use to slow the water from leaving your garden, the better the site can withstand drought. When you have a rainy period, it is also good to control the water flow.

Different sites have different conditions. On a steep site, it is about preventing gravity from transporting the water too fast. This can be done by – instead of letting it run straight down – steering the flow of water along the slope and possibly making a smaller pond that halts the water on its way down. If there is a fair bit of water and a sufficient slope, the energy of the water can also be utilized.

On a flat site, however, it is a matter of finding ways to gain momentum of the water. This might even mean pumping the water to achieve a natural flow. For this, the water is pumped up into a raised tank of some type, from where it can be released down to the crops. This may seem unnecessary; you should just be able to pump the water directly out to the crops, right? No, because the plants will be happier with room-temperature water, so it is better to have it sit in a tank first. Also, water that moves across and through the soil will pick up lots of nutrients that are not present when coming straight from a hose.

Swales halt and collect water

A swale ditch is a combined cultivation and irrigation system that not only efficiently takes care of the groundwater, but also benefits the crops along its edges.

Swales are most commonly used on larger sites. Still, the technology can also be used on a small scale with a single swale where the groundwater is captured, to the benefit of bushes and trees.

A swale is easiest described as a ditch running along the site's contour lines. Parallel to and on the lower side of the ditch is the raised cultivation beds. An example is found at the end of this chapter.

If the site has areas where it is not possible to dig ditches – like mountainous areas – a solution may be to transport the water past these parts via pipes.

Before starting work on a swale ditch, you need to review your specific conditions and decide where the waterline should run.

It may also be appropriate to make one or several ponds along the waterline. These might be ponds for irrigation, to water your animals, or just for pleasure's sake.

The distance between each swale ditch depends on how much rain you get. A rule of thumb is that the distance should be 5–6 meters for rainfall of about 1200 millimetres per year and about 15 meters for rainfall of 380 millimetres per year. Similarly, the swale ditches themselves need to be fashioned according to how much water they will hold.

A smaller swale

Begin by planning where the ditch should be placed, and if there is something that needs to be addressed before you start creating the swale ditch.

During the excavation, it is crucial to ensure that the bottom of the ditch has a slope so that the water flows in the right direction. At the same time, overflow drainages must be created, both at regular intervals along the entire ditch and the end of each ditch. These overflows are needed to protect the system from being destroyed by heavy rain or large amounts of water from melting snow.

You should avoid walking on the bottom of the ditch while working on it so that the soil there will retain its permeability.

→ *Start by marking out the width of the ditch, which should be at least 30 centimetres.*

→ *Cover the surface of your future cultivation bed with a thick layer of cardboard. Feel free to let the cardboard stretch a little outside the dimensions of the finished bed. The bed should be twice as wide as the ditch.*

→ *Dig away the turf from where you marked out the swale ditch, and place this upside down on the cardboard. This can be done either by hand or with a small excavator.*

→ *Dig the ditch down to 30–40 centimetres deep and*

Swale

place the soil you remove on top of the turf. The swale ditch is now complete.

→ *You can supplement your cultivation bed with topsoil or finely divided compost soil, if necessary.*

→ *Finish by raking through the bed and shaping it into a soft hill.*

It is good to let the construction of the ditches take some time so that you get to see that all parts of the system work as intended before you start planting.

With time, the underground water plume grows further down the ground. When it meets bedrock, the water can return to the surface in the shape of a pond or ditch.

The best plant combination in the cultivation beds is different types of trees, shrubs, and ground cover plants. With proper planning, you can potentially harvest at every level, if you want to. When the beds are new, it may sometimes be advisable to temporarily place some smaller and more fast-growing trees or shrubs that will quickly help to bind the soil.

These plants are then removed once the remaining plants have established themselves.

If you feel that an open ditch sullies the environment, you can fill the ditch with straw: It will still do its job. After a few years, when the straw has decayed, you need to replace it with new straw. The old straw is then placed

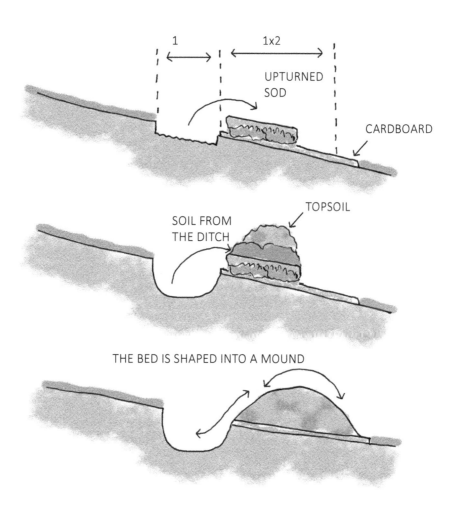

Making a swale

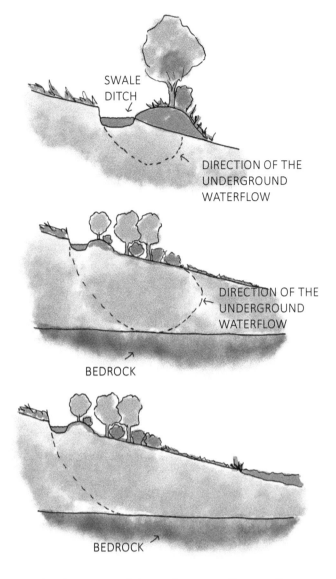

The direction of the underground waterflow

on the cultivation bed, which continues to make use of it.

Once the entire system is complete, it will be a superb resource that not only efficiently collects the water and creates a favourable cultivation climate in the beds – but which also contributes to increased control of the water's progress down the slope.

Moving groundwater

If you only need to move groundwater from A to B, creating a small brook is a beautiful way to move the water from one place to another without the use of swale ditches and other irrigation systems. Start by digging a furrow that is then covered with a pond liner wide enough to cover the edges. Then, fill the brook with natural stones, both large and small. Now simply let the water flow and enjoy your very own babbling brook.

Ponds

In principle, ponds can look any way you want them to – large or small, designed in store-bought moulds or natural shapes. But before choosing the type of pond, there are some essential things to consider.

A pond should be located where it can collect the most water. The pond should also be high up on the site, preferably above the dwelling. This is important to be able to use gravity to move the water down the site. Nature, however, doesn't always agree. If the best spot for a pond is at the bottom of the site, with houses and gardens above, it can be both resource-consuming and challenging to build a pond at the top. It is then better to leave the pond at the bottom and pump up the water to tanks placed at the top. The most important thing is that the

An artificial brook

total energy cost is considered so that you can eventually get a return for your investment.

For the plants to thrive in the new pond, it has to have at least 5–8 hours of sunshine per day. It is also good to avoid placing the pond directly adjacent to deciduous trees, to reduce debris.

Before you start digging, you need to decide how and where the excavated earth goes – it is always important to think about utilizing everything in the best way, and to save energy.

Depending on the land layer below the soil, it can be possible to create a natural pond without liners. For such a natural solution to work, the soil needs to made up from at least 30 per cent clay. An initial soil test can be performed by taking a plastic bottle and cutting off the bottom. Then, you take the soil and plug the neck of the bottle and fill the bottle with water via the open bottom, to see if it leaks. It is still wise to do a professional soil test before you start, to know for sure whether a natural pond can be created there, or if a liner or some other method must be used.

Should the natural pond still leak, there are some ways to seal it. One way is to add a layer of bentonite, which is clay used as a seal in water-carrying parts in tunnels, in the final storage of nuclear fuel, in cat sand, in the clarifying of wines, and as an anti-caking agent (E558). As long as the bentonite is wet, it will keep a tight seal, but if it gets above the surface of the water, it dries out and cracks. If you are unsure what to do, it is always best to ask a pond specialist. If you want, you can get ducks, which is a significantly cheaper and pretty sweet way to seal the pond – albeit slower.

Simply put, their poop seals the leaks over time. Still, you need patience before you know for sure if your pond is watertight since it takes about 7 years for a natural pond to saturate thoroughly. It is worth it, though, when the pond is finally linked to the ecosystem of the site.

Should you need to refill the pond for some reason, it is always best to use natural water. If there is only municipal water is available, it can cause problems for both plants and animals, since municipal water contains chemicals such as chlorine, which aquatic plants do not like. There are some "counter-chemicals" that can help, but consult with an expert before using them.

The best time to build a pond

The best time to build a pond is in the fall. The soil is wetter, so it is easier to dig. Once the pond is constructed, then rain and snow can help fill it until next spring. When spring arrives, micro-life in the pond can be kick-started by retrieving some buckets of lake water and dirt from any freshwater lake nearby.

The water level in a natural pond always moves up and down as the weather and season change. If any part of the pond is at risk of drying out during a hot summer – be careful with what kind of water you fill it with. As previously mentioned, municipal water is not recommended as it will destroy the micro life of the pond. Well water without additional chemicals work well, but it is best to use collected rainwater.

Pond depth

How deep a pond needs to depend on what works on that particular site, and whether you want to be able to keep fish. Ideally, you want your pond to have several different depths suitable for different plants, frogfish, small fish, and large fish. Just as on land, variation provides increased diversity. A deep area is not only helpful for the fish on a hot summer day but vital to whether the fish can survive the winter.

Pond shape

In permaculture, you always try to maximize the length of the edges, and when it comes to ponds; a wavy perimeter provides more edge than a round or a square. This, in turn, means more space for plants, which in turn provide more food and protection for the animals in the pond.

Plants and the environment in ponds

Similar approaches apply to pond plants; choose different varieties for different types of insects, and preferably plants that are edible to us humans. It may be a good idea to start with easy plants first, and then as you get more comfortable, supplement with a little more difficult-growing plants such as water lilies. The plants

Different pond shapes

that are easiest to place in a new pond are the ones that you can buy pre-planted in baskets. You can also choose local varieties here as a base to make sure the pond gets up and running. Organically developed plants mean that you do not introduce chemicals to the pond. One tip is to choose tall plants around the edges, to shade the water. This reduces both algae growth and provides shade and protection for aquatic animals.

To make the pond even more natural, you can add stones of different sizes, both along the edges and at the bottom. Preferably also some branches and an old log. The more natural the environment, the more wildlife will be attracted.

Fish in ponds

For fish, the same applies as when starting with plants – start with the easygoing types and try new varieties later. There are quite a few fish suppliers, so it may be wise to ask around to make sure you get healthy and viable fish. Depending on where you live, there are also different edible sorts of fish you can raise.

Algae issues in ponds

Most ponds will, at least temporarily, have algae growth. The problem is usually prevalent in spring and is due to a biological imbalance as the plants have not established themselves properly. At the same time, the water surface is exposed to intense sunlight.

Another important reason may be that the pH of the pond is too high. The ideal level is a pH of around 7.0. A pH of more than 7.5 is usually considered too high in these contexts. The best solution to algae problems is to have enough plants and, as previously said, to help nature along the way by grafting a natural ecosystem from any nearby lake. A little patience can be needed in the first few years while the pond establishes itself.

Fences around ponds

If you have small children or live in densely populated areas, there may be requirements for fences around the pond. Check the rules of your municipality. If you do need a fence – the best fence is one made from regular chicken wire. This both stops the children and gives the plants free reign, while small animals can pass through the gaps.

The benefits of ponds

The benefits of ponds are many. They provide an increased ability to store water, and the climate around the pond will be milder, thanks to the increased solar radiation. Ponds also attract wildlife of all kinds, which contribute in different ways to the whole area. In fact, it only takes

a couple of days before the insects start to appear. With time, you also get a lot of additional composting material when the vegetation in the pond needs to be thinned out and pruned.

We humans, like all animals, are also drawn to the water. It's good for us to have some water at home.

Aquaculture and hydroponics

Aquaponics, or aquaculture, is a system where you grow plants and fish in the same system. A system where you grow plants (without fish) in water is called hydroponics.

Aquaponics is perceived by most as something new, but the technology comes from nature itself. In a large natural pond, the entire natural system of fish and edible plants in, around, and on the water is one single extensive aquaculture. If the pond is too small to grow edible fish in, you can always grow small fish for bait to use when fishing nearby.

Both aquaponics and hydroponics are available in anything from large systems to small ones that you can keep indoors. You can now even buy small, ready-to-use systems. Although it is cheaper – and quite simple – to build a system yourself. Practical instructions can be found online.

Pumps

The type of pump that is the best fit for any given situation is determined by location and plans. The questions are many: Should the pump just oxygenate the water, or should it run a filter or a fountain? Should it move water along to a brook? Is it sufficient with one pump, or do you need several? And so on. It is vital to go through all the details – and if you feel uncertain, it is wise to seek help from someone knowledgeable.

Here are some tips to get you started. If it concerns a small pond of less than 6 sqm, then one pump is usually enough. A basic rule is that the pump should theoretically be able to circulate the entire volume of water every two hours.

The rule of thumb for a slow, babbling brook with a width of 10–15 cm is that a water flow of 1500–2000 litres per hour is needed at the outlet. If you want a rushing brook of 20–25 centimetres wide, this equates to 3500–4000 litres per hour. In a circulating system, one must also take into account how high the water must be lifted to get back to the pond, which in trade terms is called lift.

If instead, it concerns a pump solely for pumping up water to a higher pond, lift (sometimes also called head height) is the deciding factor. All pumps need energy, and the most common solution is electricity – but there are more and more alternative solutions on the market. You can even get smaller solar-powered pumps in stores right now.

Hydraulic ram

Lastly, I would like to take the opportunity to bring up the tried and tested hydraulic ram – also known as a hydram – which was invented in the late 18th century. It is a simple design that pumps up water with running water as the only source of energy. The disadvantage is that only a small part of the water coming in is pumped up, which means that it may take some time to pump large amounts of water. But with a tank at the other end, this is both an inexpensive and reliable way to use nature's water. If you do not want to buy one, you can build a hydraulic ram yourself using one of the many instructions available online.

Ängsvägens water

The existing water supply consists of well-water from a private well, and rain barrels under two of the drain pipes. For more extended power outages there is a communal hand pump available, as well as a lake 300 meters from the house to collect water from. Neither of these sources provides drinkable water. The nearest place to collect drinking water is a few kilometres away.

The site is very hilly and therefore mostly dry – except on the large meadow at the bottom. The ground here is quite waterlogged during a typical summer, which is partly due to the immense water pressure from the surrounding groundwater and partly since the greywater infiltration ends in the northern corner of the meadow.

The solution to the water issues at Ängsvägen

Rain and condensation water

The very top of the site is mountainous with no possibility of collecting rainwater. The first level from which rainwater can be collected is from the roof of the house. There are already gutters and pipes that collect the roof water. Still, the system needs to be expanded with more water barrels and tanks that hold a larger amount of water.

On the back of the house is an air source heat pump

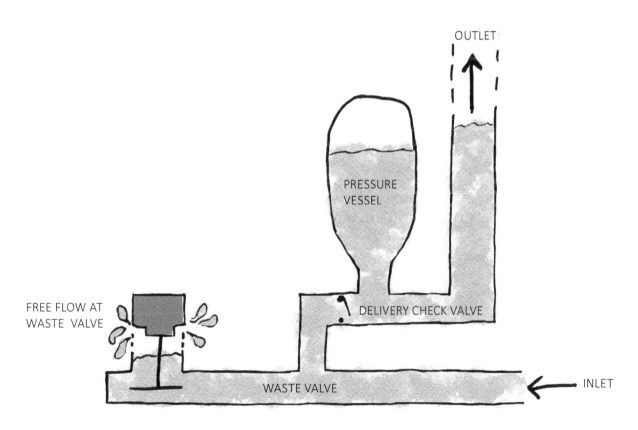

Hydralic ram

that produces quite a lot of water from condensation that is not utilized. Pipes should transfer this water to a larger tank of 1000 litres at the northwest corner of the house. The rain and condensation water will primarily be used for pot-grown plants outdoors, and the greenhouse.

Swales and ponds

Thanks to the site's hilly appearance, it is ideally suited for a system based on swales.

The starting point of the system is placed in such a way that all groundwater coming from the mountainous area above is effectively collected. Halfway down the slope, the ground flattens out. There it is suitable to build a smaller pond for the family's future ducks.

The water then proceeds down from there to the northwest corner of the meadow. Down here, the system's largest pond is placed to collect any and all water. The goal when creating this pond is to design it as a small, natural lake; with some depth for the family to grow fish here.

This assuming that the soil sample shows that it is possible. The overflow drainage then leads the water to the existing ditch along the road – which eventually moves the water to the site's last pond.

In this last little pond, it is advisable to cultivate different varieties of reeds in order to utilize all of the nutrition in the water. The reed is then cut down at regular intervals and is either used as mulching on the garden beds further up in the system or added to the compost.

The largest pond will also supply the large cultivation beds on the meadow with water. During a typical summer, these beds won't require much water – but if it is hot and dry, they will require regular watering. The best solution for this is an irrigation system that is operated with a pump that is connected to the drip hoses in the –respective bed. Preferably a solar-powered pump. The irrigation system is taken down during the winter.

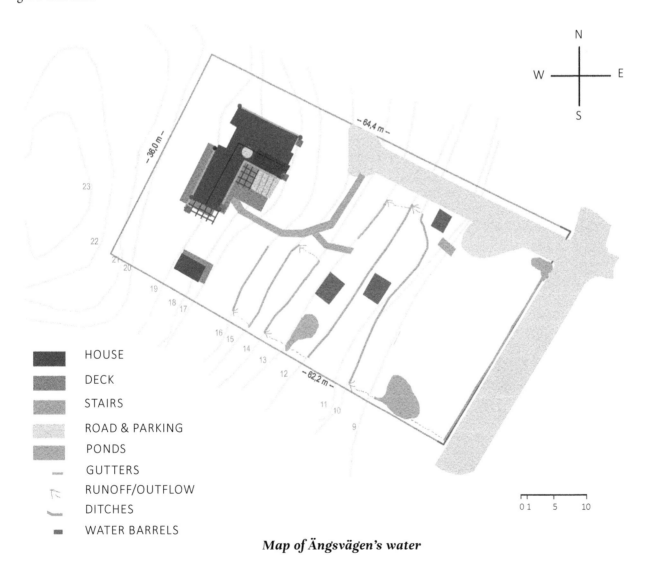

HOUSE

DECK

STAIRS

ROAD & PARKING

PONDS

GUTTERS

RUNOFF/OUTFLOW

DITCHES

WATER BARRELS

Map of Ängsvägen's water

To understand and grow the soil

Learning how your soil works can take some time, but you can learn a lot by merely observing the ground itself and what naturally grows there. Those "pesky" things we usually call weeds are in reality an encyclopedia, where many types of weeds act as a signaling system that tells you how the soil is feeling. When the ground is imbalanced, these "weeds" attack the problem in that particular area. In fact, there can be up to 2000 dormant seeds on one square meter of land. Seeds that are there, just waiting for "right" conditions to arise, so that they can start to grow.

Different soil types

Different soil types retain varying amounts of nutrients and water.

Sandy

Sandy soil is easy to work and is relatively airy thanks to its large grain size. That is why various root vegetables grow large and straight in sandy soil. The disadvantage is that this type of soil is poor at retaining both water and nutrients. Sandy soil heats up faster in the spring but also gets colder in the autumn.

To increase the humus content in the soil so that it is better at retaining nutrients and water, you add – for example – grass clippings, manure, and compost. Even mulching benefits water retention.

Fine Sand

A soil made up of very fine sand has properties similar to the sandy soil, with the difference that this type of soil is better at retaining water. Fine sandy soil can be improved in the same way as sandy soils.

Coarse Silt

Coarse silty soil differs a great deal from the fine sandy soil. The structure is grainy and is even better at retaining water. The disadvantage of the grainy structure is that it easily breaks down; for example in the event of heavy rain, and then forms such a hard surface that the seeds that grow cannot penetrate the surface.

Coarse silty soil is improved in the same way as clay soil.

Silt

Silty soil is a soil that is good at retaining nutrients and water. However, sometimes this soil can be so water-retaining that it causes issues. The soil becomes wet, compact, and above all – cold, which means that this silty soil takes longer to heat up in the spring. Compact and wet soil is also easily affected by a lack of oxygen, which causes the plants to suffocate. Silty soil is improved in the same way as clay soil.

Clay

The clay soil is the finest of the various soils. It is a very nutritious soil, but with the disadvantage that it is wet and cold. In addition, the clay soil is especially sensitive to soil compaction.

Improving clay soil, coarse silty soil, and silty soil is done by adding large amounts of organic material, by mulching, and by growing green manure plants in order to loosen up the soil and get its' micro life started. Do not add sand to clay soil. This can make the clay soil coalesce even more and become as hard as concrete.

Loam

Loam is a mixture of sand, silt, and clay in different amounts, so, therefore, you need to try and determine which of the above soils it most closely resembles. Adding organic material always works.

The grain sizes of the different soils

Soil type	Grain size
Gravel, medium	20–6 mm
Gravel, fine	6–2 mm
Sand, coarse	2–0,6 mm
Sand, medium	0,6–0,2 mm
Sand, fine	0,2–0,06 mm
Silt, coarse	0,06–0,02 mm
Silt, medium	0,02–0,006 mm
Silt, fine	0,006–0,002 mm
Clay	0,002 mm

Testing the soil to figure out the type
Feel test

Moisten the soil and rub it between your fingers. Sandy soils (e.g. sandy loam or sandy clay) feels gritty, silty soils (e.g. silty loam or silty clay) feels smooth, and clayey soils (e.g. clay loam) feels sticky.

Ball test

Roll a ball of approximately 25 g moistened soil in your hand.

Coarse textures (sand or sandy loam) won't ball at all, or the ball breaks at the slightest pressure.

Sandy loams and silt loams stay together but change their shape easily when pressing them.

Fine-textured (clayey or clayey loam) soils stay together very well.

Ribbon test

Using that ball you just made, press it out between your thumb and the side of your index finger in an upward motion so that a ribbon forms.

Sandy soils won't ribbon at all.

Loam, silt, silty clay loam, or clay loam soil ribbons less than 2.5 cm before breaking off

Sandy clay loam, silty clay loam, or clay loam ribbons 2.5–5 cm before breaking off

Sandy clay, silty clay, or clay soil ribbons more than 5 cm before breaking off

Sound test

If you take some dry soil and rub it between your fingers and it crackles; the soil is mostly sand. If it instead makes a creaky sound, the soil is made up of finer particles.

Test pit

One way to test the soil's ability to retain water is to dig a test pit that is 50 centimetres wide and 30 centimetres deep – and then fill it with water. If the water remains for several hours or even worse; several days, the drainage is too weak, and the soil is too good at retaining water. If you want to plant something there, the soil must be improved. It is easiest to do this by mixing in more organic material such as compost, grass clippings, and so on. If, on the other hand, the water disappears within 30 minutes, the drainage and the water-retaining ability of the soil is good. If the water disappears within 10–15 minutes, the soil is porous, which is also addressed by organic matter.

Dig test

When digging in good soil, it should be possible to push the spade down into the ground solely with the force of your foot, and the soil you dig up should fall apart. In a soil that is too compact, it is harder to push down the spade, and the soil comes up in clods. To make the soil more porous, organic matter is added, preferably in combination with the cultivation of deep-rooted green manure plants.

Water test

The easiest way to test how much clay soil contains is to put some soil in a jar of water and then shake it rigorously. After a while, the different soil types will settle depending on their size. Sand in the bottom, then sludge, then clay and at the top, organic material.

I am convinced that by now, no one has missed the fantastic properties of organic material, right?

The art of watering

Back again to the subject of water. Depending on the type of soil you have, you should water it in different ways. A sandy soil that is poor at retaining water should be watered frequently and a little at a time, in dry weather. When it comes to a soil that is mostly clay, you water rarely but a lot at a time – because large amounts of water are needed to penetrate the dense clay.

Regardless of the soil type, it is good to add humus as it balances the soil's water retention. Sandy soil with a high humus content does not need to be watered as often, while a clay soil with a high humus content needs to be watered more often.

An average water requirement is about 30l/week, but your kitchen plants need more. Some need as much as 200l/week. The comparison should be that 1 mm corresponds to 1l/m2. Therefore, you need at least 30mm/week, and up to 200mm/week for some plants. That is why the art of watering is so important.

I have mentioned this before, but it bears repeating. You should not water between 9 am and 4 pm (solar time) when it is hot out. Watering during the hottest hours in the belief that you are cooling the plants down, in reality only gives the plants a shockingly cold shower.

Watering in the morning is best – although watering in the evenings usually works as well; at least if you do not have a problem with snails or plants that are easily attacked by mould. In the greenhouse, watering in the evenings should be avoided as far as possible to prevent mould.

An old and well-proven trick for optimizing watering for perennials requires, first and foremost, them to be planted slightly higher than ground height; on a small hill. Thanks to this planting method, one can optimize the plant's access to water by making a small pit in the hill during spring, in order for the plant to absorb the water properly. When you reach July, you water as little as possible. In August, you round off the soil so that the hole is gone and you have a proper hill again. This means that the water flows off more quickly and that the plant settles better during autumn, and thus will cope better in the winter.

It should also be remembered that established perennials have more extensive and deeper root systems, which means that they need to be watered less than one- and two-year-old plants. Newly planted perennials, on the other hand, need to be watered more often and with large amounts each time in order for the root system to develop and spread down into the ground. Mulching helps to keep water consumption down, which landscape fabrics also do.

At high winds, the evaporation rate increases and you get a definite increase in water consumption if using irrigation, because the light water droplets drift away in the wind. An effective way to keep water consumption down is to use a drip irrigation system.

Another factor that affects water consumption, which one might not think of, is when the soil has become dry and hard before watering. Such a soil surface has a much harder time absorbing the water, and you lose a significant amount that trickles off somewhere else, at the same time as soil erosion increases.

Mulching

Covering the bare soil around your planted crops has many advantages and few disadvantages. When the plants grow to such an extent that they are large enough not to risk drowning in the cover material, it is time to weed and start covering the crops. What to cover with depends a little on what you have access to.

Comfrey is considered the very best, but if you are to grow it yourself, you need to be aware that comfrey is a so-called invasive plant, which means that it quickly takes over the area you plant it in. The least invasive variety is said to be the Russian comfrey plant, (Symphytum uplandica = Symphytum x uplandicum). Alternatively, one can plant some form of green manure plant between the rows, which are then cut down and used as cover material. More about green manure later on.

Advantages

→ *Prevents soil erosion during high winds and heavy rains.*

→ *Improves the soil structure by decomposition, which in turn increases the soil's humus content.*

→ *Stimulates the soil's microorganisms, especially the mycorrhizae.*

→ *Increased mycorrhizae improve the growth and uptake of many micronutrients while protecting the plants from high levels of harmful substances in the soil and increasing the plant's resistance to diseases.*

→ *Reduces evaporation and helps the soil retain moisture both longer and in a larger volume. A thatch of as little as 3.5 cm reduces evaporation by about 35 per cent, compared to bare soil.*

→ *Stabilizes the soil temperature in the topsoil layers, which means that the smaller fine roots have a higher chance of survival. With bare soil, even established plants can suffer from stress because the fine roots do not like to be exposed to large fluctuations between hot and cold.*

Mound

Donut shaped mound

→ *Prevents silting of the soil surface during heavy rains. Suffocates the weeds.*

→ *Legumes (including vetch, fodder peas, and clover) in particular increase the number of nutrients in the soil.*

→ *Requires less weeding, nutrition, and watering.*

→ *Greater yields.*

Disadvantages

→ *Increased snail problems in areas with larger snail populations.*

→ *Can be counteracted by growing in raised beds with some form of snail barrier around.*

Various cover materials

Cover material should primarily be organic to gain full benefit from the mulching.

Grass clippings provide quick nourishment and are suitable for nutritious plants such as leaf plants, cabbage, tomato, cucumber, and leeks.

Straw, hay, and leaves are broken down more slowly. They are well suited for plants that require fewer nutrients such as potatoes, root vegetables, onions, peas, and beans. These also suitable for mulching around berry bushes and fruit trees, because it helps keep the plants healthy and increase the harvest.

Contrary to previous data, bark mulch and sawdust are now considered to have a positive effect on the general nutritional content of the soil. Yes, even woodchips from conifers work in some cases as cover material. The condition for this type of woodchips to work positively is that it does not melt into the soil. It must, therefore, decompose above ground so as not to acidify the soil. Some data also show that soil covered with chips from coniferous trees can raise its' pH. The nitrogen deficiency that arises in the soil surface just below the chips makes it harder for the seed to grow properly, which on the one hand gives fewer weeds, but on the other hand, does not work when sowing directly on the open.

Remove or wait to add your cover material – mulch – until the seed has sprouted and the plants have established themselves. If you have a heavy soil such as clay soil, you need to remove the mulching in the spring so that the soil warms up faster.

Weeds are not the issue

The solution to your "weed problems" is not to remove the weeds, not without correcting the underlying issue. Instead, you have to consider why that particular plant grows there. Compare it to you taking a pill because you have a headache – it is hardly because the body lacks acetylsalicylic acid, the cause is something completely different. The same goes for nature.

Depending on the way the soil has been handled, nature responds with some particular seeds growing to repair the damage.

In the appendix of the book, I have compiled a list of plant indicators which I hope you will enjoy.

A fertile soil

Fertile soil is the basis of everything. First, there must be enough nutrients for the soil to maintain its fertility; then there must exist a surplus that can be used to grow with. This means that the soil will gradually be depleted if you remove more nutrients than you return to it. An easy way to keep some nutrition in when harvesting is to let the plant parts that you do not want, remain on the ground to decompose.

Different soil types and their weeds

Land type	Plant type
Tightly packed, no air	Deep-rooted plants (taproots) helps to loosen the soil, such as thistles
Ploughed many times, very airy	Small bush-like plants with a hair-like network of roots
Burned	Leafy grass and ferns are utilizing the potash (potassium carbonate) formed in the ground after a fire. If you continue to burn the ground, these will spread. Instead, if you instead cut them down and mulch them, the area will decrease its production of grass and ferns.
Wrongly used for a long time, i.e. overused	Herbaceous plants with flowers that form pods, and whose roots have small tubers of nitrogen-fixing bacteria.

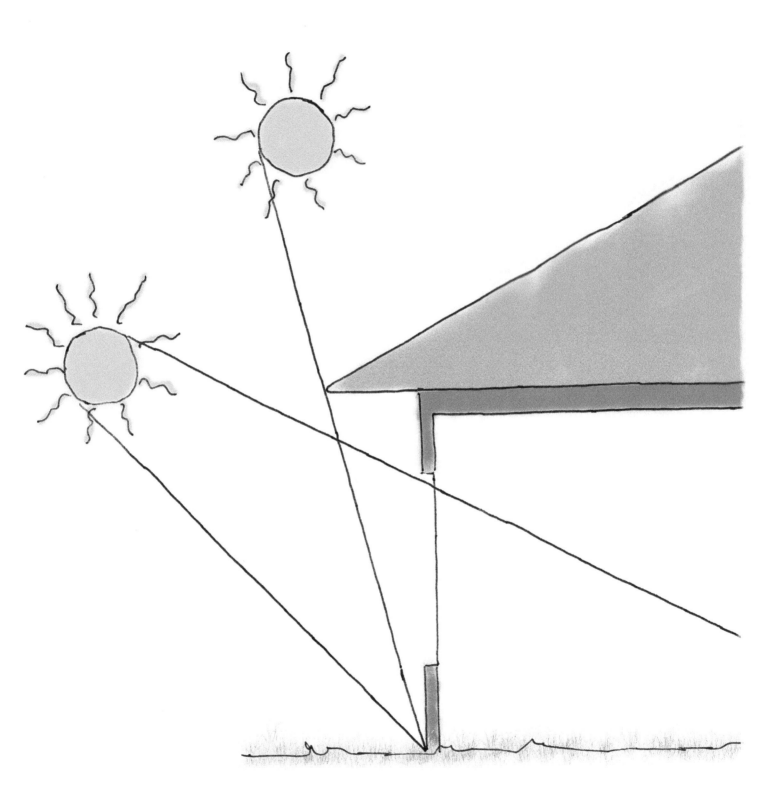

How climate affects the need for energy

According to many, the climate in Sweden is perfect for a permacultural way of life – as long as we can satisfy our need for heat and storage.

The conditions of a place

When working with permaculture, and when it comes to energy and climate – you plan to utilize the incoming resources to the maximum before they leave your yard while creating conditions to protect family and home from various climate and energy challenges. To gain knowledge of the conditions of a place; the influence of water, precipitation, wind, and sunlight on the land and buildings is studied. With the combined information, you can then use the permaculture design tool to optimize both energy usage and energy storage. The ultimate goal of optimizing energy resources is – at least in the long term – to be self-sufficient and, if possible, to produce a surplus for any emergencies.

Energy from water

Water can also be a source of energy within a system in various ways. According to the laws of physics, the higher up in a system the water is, the greater the amount of energy it contains. On the other side of the coin, the farther down the system the water is, the less power the water has – while also increasing the potential for life in it.

If you compare a roaring waterfall and the still lagoon at the end of it, it is easier to understand the connections.

The influence of the wind

The wind affects the climate a whole lot. By observing different wind phenomena over a long period – at least one year and preferably several – you can determine whether wind protection needs to be created. If you need to open up for more wind, or if there is any possibility of taking advantage of wind energy.

The sun and our homes

The sun is the largest energy source we have, and it gives us vital light and heat. In fact, without the sun, there would be absolutely nothing, but today – the sun has come to represent not only an asset but also a threat. In climate change contexts, the effect of the sun is often discussed as something that we must protect ourselves from, to avoid an even greater issue. 'This can create a vicious circle, which has already happened in several places around the globe as the amount of smog and cloudiness grows, decreasing the possibility of natural farming and gardening due to a lack of sun.

When building houses today, there is much to gain from taking into account the sun and its energy. Here in the Nordic countries, it is especially important to orientate the houses towards the south. The best energy result gives a location where the long side of the house is slightly turned towards the southwest because then the last rays of the evening sun will also benefit you. For us in the north, the best house shape is a long, narrow one.

Letting the room planning be dictated by the sun is also beneficial from an energy perspective. This is because the best location for bedrooms, bathrooms, toilets, laundry, and mudrooms are against the shaded side of the house. In contrast, the kitchen, dining room, living room, and other day rooms are best placed against the sunny side.

The windows should also be planned with regard to the sun, where the ones on the shaded side are few and quite small, while those on the sunny side are many and large.

The fact that the sun hits differently in different seasons may not always be something you think about. Still, in regards to energy, this is important. In the middle of summer, the sun does not reach far into a room, but it often becomes too hot indoors anyway. From an energy-efficient point of view, it is better to use some form of shading to protect you during the hottest hours of the day, instead of using energy-intensive air conditioning. In the autumn; and in line with the ever-sinking sun, the sun reaches farther into the house, which can be used in various ways.

If you are building from scratch, the walls that are reached by the winter sun – most often the middle wall of the house – can be made from a material that has an excellent ability to store heat. The winter sun will then heat the interior wall, which then slowly emits its heat during evening and night. If you want even more winter sun, a good solution is a split roof as in the picture on the next page.

In an existing house, you have to resort to other tricks such as having these walls darker or using different forms of interior design that can help to preserve the heat.

I want to take this opportunity to recommend a relatively simple, cheap, and above all, solar-powered air conditioning unit – a so-called sun chimney, which is based on the principle that hot air rises. A sun chimney consists mainly of black plate tubing, a roof passage, and

Different roof heights for more sun intake

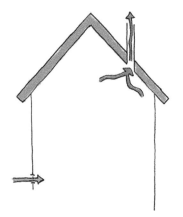

Solar chimney

a vent inside the house. The chimney should be located on the south side of the roof, right above the warmest point you have in your house. When the sun hits the tube, this becomes hot – very hot, which causes the air inside the tube to rise, while at the same time pulling on the warm indoor air. To get good airflow, this is supplemented by a few vents along the floor of the shaded side of the house so that the cooler air can be drawn in. It is important to remember that all vents must be able to close entirely to avoid heat losses during the cold season.

An excellent alternative to harness energy is to connect the sunny side of the house with a greenhouse. To get the most out of the heat this so far north, a rather large part of the house wall is covered with windows. The rule of thumb is that the sunny side must be glass-covered with as many percents as the latitude you live on – if you live at a latitude of 60, The wall's surface needs to be 60 per cent glass. If this is not feasible, an added greenhouse is still better than nothing; since in any case, it reduces the heat loss of the wall. More about this in the chapter "Extending the season".

Finally a little about different fireplaces. The colder the climate, the more critical it is to have several fireplaces, preferably placed along the middle wall of the house. In addition to the usually tiled stoves in Sweden, there are several other solutions such as the German "Kachel offen", or the Chinese "Kang", which often

Rocket stove/Rocket mass heater

also serves as a bed. This "Kang" can be compared to a so-called "Rocket Stove" both of which can be likened to a horizontally tiled stove.

A Rocket Stove is very effective, especially when using with branches and sticks as firewood. A 30-minute fire can, if you have proper construction, provide heat to an entire house.

In fact, many people think that using regular firewood in tiled stoves and the like is a pure waste of resources – because branches and twigs have a better heating effect.

Also, branches are often something you have left "over", in many different contexts.

Microclimates

Understanding how different microclimates work provides many advantages. Their functions not only give us opportunities for better growing results, but we can, in general, create a better place to live.

The role of topography in the climate

Topography is heights, valleys, and slopes. When planning a garden, it is easy to just think of the weather, whether there is sun or shade, and if it is windy. But the topography is just as important.

Sun angle

The sun's angle to the earth is crucial for the climate and what you can grow. Thus, the slope of the ground in a specific location also plays a vital role in how large the solar radiation becomes.

If a slope leans towards the south, each degree increase means a higher solar angle with the corresponding number of degrees. The sun is naturally higher in the sky on a southern slope, which also means that the shadows become shorter. Conversely, if the slope is instead to the north, then the shadows grow longer and the sun angle lower, the greater the slope. A slope can, therefore, affect the climate, several zones within a small area.

Depressions

In everything from small dips to vast valleys, the risk of frost is more significant than on the ground above as cold air travels downwards. The cold air drops to the bottom and remains there. Even in the tiniest little depression, the cold air remains and can cause frost, especially if there is no wind.

Wind and water

The steeper the slope is, the faster the wind moves up the hill, which in turn creates an increase in turbulence on the leeward side of the hill. If the slope is not too big; a well-placed shelter can eliminate the risk of turbulence. If the slope is too steep for wind protection to help

protect the leeward side, the wind's energy can instead be used in a wind turbine.

Water flows, as you know, faster the steeper the slope. One effect of this is that the faster the water flows, the shorter the time the surrounding soil will have to absorb it. At the same time, the risk of soil erosion increases.

The soil's condition affects the climate

The soil influences the microclimate primarily through its ability to hold water, and through how much water evaporates from it. A soil that binds more water provides a warmer and more humid microclimate. After heavy rain, the effect can be so significant that it corresponds to the impact of an adjacent pond or lake.

If the soil is bare or if you are mulching also affects the microclimate because the bare soil reflects more light and heat than the covered soil does.

Water provides a milder climate

The climate in one area is very much affected by if there are lakes around, as well as if the site is located by the sea. Since water is heated and cooled down slower than air, the proximity to surface water provides a milder climate with less severe temperature fluctuations. The evaporation of water also affects the climate. This effect can be exploited on your site by constructing various forms of open water collectors. An example is if you have a dip or hollow in the landscape where frost easily forms. Then, by creating a pond in that place, it is not only possible to prevent frost, but also to positively affect the immediate surroundings.

Next to any vegetation, rain falls unevenly depending on where the wind blows. On the windfall side you will have a wetter climate, and on the other side ends up in a so-called rain shadow.

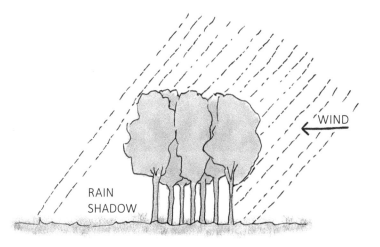

Rain shadow

Plants regulate temperature

The vegetation in a permacultural system is planned so that it interacts in various ways to create the best possible microclimate. The plants not only cover the ground and prevent heat loss, but they affect most things. The plants regulate the temperature both above and below ground, they filter dust and other particles in the air, and they act as windshields and as shadows. If you compare the climate of an open area with that of a place next to a tree with large leaves, the difference is highly noticeable.

Why; there are even plants that are their own heat source. One example is crocuses, that can heat themselves 15 degrees over the surrounding air temperature. It is not so strange then that these little fellows are often the first to show themselves in spring – even through the snow!

Houses and other constructions collect heat

Our homes also affect the microclimate by absorbing heat during the day and delivering it at night. By, for example, creating a windshield that reduces harsh winds on the house, the house can absorb more heat, which is then released more slowly.

All forms of "artificial" constructions can be planned to maximize the opportunities for creating pleasant microclimates. This can refer to patios, hard surfaces, fences, walls, and stones, for example. A simple thing like placing stones strategically next to a cold-sensitive plant can make it grow and thrive.

House location

If one is to rebuild, the key is to carefully study the landscape topographically before deciding where to build. If it is a sloping plot, the best position is around the middle of the hill. This is to both avoid harsh winds at the top, frost at the bottom, and the possibility of utilizing the earth's gravitational force when the water system is planned. A house at the top is directly counterproductive because everything that needs to be moved up requires a lot of energy. The only advantage is a possible view. Then, it is better to make a beautiful lookout up there instead. Also, the temperature around the house is affected by its position, because it is colder the higher you get.

Frost

Frost creates many problems for us up here in the north, but if you are observant and a little prepared, milder frost can be managed reasonably well. First and foremost, you need to get to know the plants, the climate, and the microclimates so that you know where the risk of frost is most significant. Then several other factors make it easier to avoid frost.

Keeping track of the temperature

Checking the forecast for the night temperature is both simple and effective. If the prognosis says 3–4 degrees, there may be a risk of local frost depending on how all the other factors affect you. The most significant risk of frost is at dawn.

Raised beds reduce the risk of frost, precisely because they are raised. Stone and water absorb heat during the day, which is emitted during the night. Stone slabs or bricks around various perennial plants can help protect against frost, to name one example. Another easy and removable trick is to add water-filled milk cartons (those with screw caps) around sensitive plants. Heat is then built up during the day and delivered at night. If you want to put some extra work into this, you can paint the cartons a dark colour. It increases heat absorption, and the cartons become more beautiful. How long the heat lasts depends on how hot the day has been, how large the amount of stone/water is, how late at night the frost comes, and how long it lasts.

The heat loss at night is proportional to the surface of the sky to which an object is exposed. In order to understand this more easily, you can compare it to a mouse sitting in a cardboard tube on the ground, looking out into the night. The mouse sees only a small, small part of the sky, which means that it is not exposed to large amounts of cold air. On the other hand, if the mouse climbs out into the open, the exposure surface is the entire sky which means that the mouse must handle a much higher degree of chilly weather. This is why you should crawl under a large tree – preferably spruce – if you have gotten lost and need to stay the night in the forest.

Clouds provide a lower risk of frost

A starry night sky quickly releases the heat of the day, while clouds effectively stop the earth's heat radiation. Therefore, the risk of night frost is much smaller on a cloudy night.

The wind can both increase and decrease the risk of frost

Most people know that the risk of frost increases if there is no wind, but just because there is wind does not automatically mean that the danger has passed. In order for the wind to protect against frost, there must exist mild winds that cause different air temperatures to mix with each other. If you have too much wind, the risk of frost increases as the warm air then blows away, leaving room for the denser cold air to drop to the ground.

Cold air travels down

If your crops and gardens are on hilly terrain, it is

especially important to remember that cold air drops and warm air rises. When the temperature falls at night, the cold air slides down the slope and lays as a lid over the lowest area. At the same time, the warm air rises along the slope. This applies both in large and small scales. As stated before, even the tiniest little dip can suffer frost due to this.

The length of the night determines the risk of frost

The longer the night lasts, the longer the heat has to flow out into space, and with that increasing the risk of frost. This is part of the explanation why there are frosty nights in the early summer and early fall, but not in the middle of summer.

Irrigation against frost

The so-called dew point is the temperature at which the air can no longer retain the moisture present in the air. This phenomenon can be used as frost protection due to when moisture condenses; it creates heat – which in turn helps to keep the frost away.

Therefore, a light watering one or two days before there is a risk of frost can help to keep the frost at bay. Only a light watering, mind – as a more substantial amount of water in the soil can instead increase the risk.

In larger ventures, it is common to use irrigation to prevent crops from freezing. By placing a thermometer about one meter above the plants, and hooking up a simple alarm that is connected to the thermometer, one can quickly know when it is time to go out and turn on the water.

Trees protect against frost

The heat that the trees store during the day protects the immediate surroundings from frost. Also, the trees' perspiration protects due to the moisture content of the air increasing, thus increasing the heat. One way to take advantage of this is to grow around and in the vicinity of, for example, fruit trees.

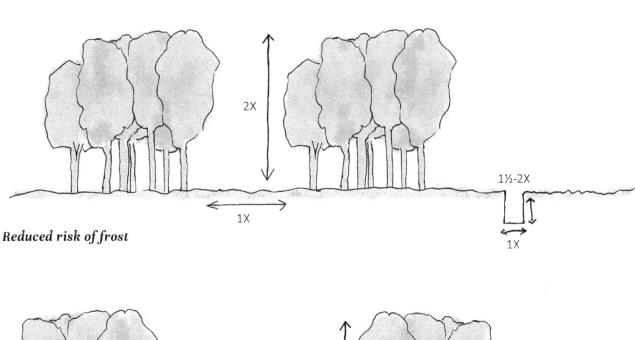

Reduced risk of frost

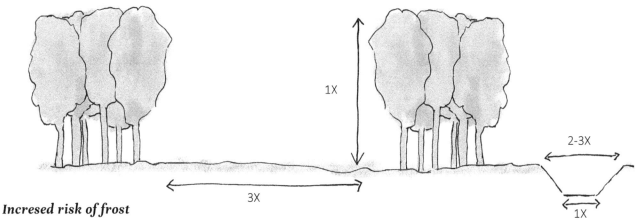

Incresed risk of frost

Coverage provides extra frost protection

Mulching itself reduces the risk of frost, at least if one compares with cultivations where the soil lies bare. But usually, this is not enough – you need to use some other form of temporary coverage. The most common is a so-called garden fabric, but curtains, sheets, and blankets also work well as long as the plants are not exposed to excessive weight.

The coverage of taller plants can be a little harder to get right because they quickly become a wind trap. Whatever the case, all coverage needs to be anchored to the ground, and for this, both stones or bags with soil work well. To counteract damage to the plant, it is good to have some form of construction on which the cover fabric rests. Creating a distance between the plant and the cover also reduces the risk of frost damage to plant parts that otherwise have direct contact with the fabric.

Wind and windbreaks

As previously explained, the winds affect the climate a whole lot. Everyone knows what a difference it is between a hot, windless summer day and a hot day when the air provides a cooling breeze. Also in the winter, a beautiful day of two degrees below zero and radiant sunshine can be turned into a day of biting cold if you get heavier winds. But the wind can, as many other things in nature, be adapted to a certain limit if only we understand and cooperate with nature and its conditions.

We humans have it the easiest when it comes to handling different wind situations – we can adjust how we dress, go indoors, or otherwise seek shelter. It is a lot harder for animals and nature itself. The wild animals can usually seek shelter on their own. Still, the domesticated animals are entirely dependent on us, ensuring that they have access to protection. Nature, on the other hand, has to stay where it is, and it is precisely why looking at where the plants can tell a lot about the different conditions of that particular place.

The wind affects more than you might think. For example, both growth and production are already decreasing at as weak winds as 7 meters per second. Then, a properly created windbreak can make a big difference. But it is essential to be observant; because it can get far too hot behind the protective barrier. When the temperature rises above 25 degrees, photosynthesis gradually decreases to finally come to a stop, which also contributes to a reduction in growth and production. At 38 degrees, most new and transplanted plants die. To counteract this, one can protect the plants with a shade net during the hottest times.

The benefits of windbreaks

Control the wind
By controlling the wind, you can change the climate of a place considerably. Cold winds can be steered away

The effect of the wind on the temperature

Wind/degrees Celsius	10	6	0	-6	-10	-16	-26	-30	-36
2 m/s	9	5	-2	-9	-14	-21	-33	-37	-44
6 m/s	7	2	-5	-13	-18	-26	-38	-44	-51
10 m/s	6	1	-7	-15	-20	-28	-41	-47	-55
14 m/s	6	0	-8	-16	-22	-30	-44	-49	-57
18 m/s	5	-1	-9	-17	-23	-31	-45	-51	-59

from sensitive plants, domestic animals, homes, and from places where we often spend time. The wind can also be adjusted to provide coolness in areas that are far too hot. You can even use the wind to help you deal with snow, by strategically placing trees and bushes in places that you know will require shovelling. Just remember to observe the winds for an extended period, due to different winds resulting in different effects.

Reduces plant damage and increases harvest
Plants exposed to wind get stressed, which in the long run increases the risk of broken branches, fewer leaves, and a smaller harvest. Also, the pollinating insects find it more challenging to do their job when it is windy. Windbreaks also prevent airborne weeds from reaching your garden beds as easily. A properly planned windbreak provides you with happier plants and a more abundant harvest.

Protects the soil
Windbreaks preserve soil moisture and reduce soil erosion. When growing, living windbreaks also contributes to better soil and reduces the risk of soil erosion during heavy rains.

Reduces evaporation
During windy weather, evaporation from both soil, plants, and surfaces of water occurs faster than when it is still. A windbreak next to a pond, for example, can reduce the water loss during hot summer days.

Saves energy
Properly placed trees and shrubs around different buildings can save a lot of energy over a year. The plants not only protect against cold winds in the winter but also provide shade during hot summer days. As the climate gets warmer, the value of that shade will increase in regards to energy as well, since it is becoming increasingly common to install heating systems that can also provide cold air in summer.

Provides compost material and wood
Windbreaks that are alive must be maintained over time so that they retain their full functionality. This is mostly done by pruning, which gives you branches that can be used as additional food for animals, be turned into wood chips, and used for composting. But over the years, one may have to replace larger trees, and then the windbreak can also provide timber and wood.

Diversifies
The diversity benefits from the fact that the plants used as windbreaks are often species that you might not otherwise have. Fruit trees are not suitable to protect against the wind – this task requires other species. The trees and shrubs also effectively catch the nutrients that come with both rain and wind, which benefits both the plants and the life in the soil. Birds and insects also thrive, which further increases the diversity when using windbreaks.

Protects animals
The animals, both domestic and wild, move instinctively to the places that are best for them. You can see this when you have winds over six meters per second because the animals seek out places that provide better shelter. Or as seen on really hot days, when the animals turn to the shade of the trees – alternatively if it is windy, out into the open to the cooling winds.

Filters air
Trees and shrubs work very well as air purifiers. The dust and any contaminants that come with the wind are effectively filtered by the plants in your windbreak.

Wind can create energy
Wind contains a lot of energy that you can take advantage of. Through a suitably placed windbreak, you can steer the wind towards a small wind turbine that can then create enough power to – for example – operate a water pump.

Disadvantages of windbreaks
There are, however, some disadvantages when it comes to windbreaks, and they are:

→ *Reduced growing area*

→ *Competition between windbreak plants and crops*

→ *Shading*

→ *Increased risk of frost, depending on the surrounding environment*

→ *Too warm during hot summer days*

Planning a windbreak
The outer edge of the windbreak is the side that catches the wind. The wind is pushed up and over the windbreak along the outer edge, at the same time as the wind speed increases, which in turn can create turbulence on the lee side (inner edge). When the wind starts to swirl behind a windbreak, it twists to the right (here in the northern hemisphere) while also being displaced 15 degrees to the right.

Windbreak profile

How the windbreak looks in profile is crucial for how much lee it can provide. An inclined, dense and extended profile means that the wind is pushed up over the canopy, with turbulent winds on the lee side as a result. However, having the outer edge as vertical as possible, with a relatively sparse planting pattern results in that part of the wind being sifted through, which in turn results in considerably less turbulence on the lee side.

Canopy surface

The appearance of the canopy also affects the wind quite strongly. An uneven canopy with trees of different heights breaks the wind stream, and thus reduces the wind speed even before it has passed the windbreak. If the canopy is flat, the wind can actually become more forceful on the other side of the windbreak, as the wind gets an extra boost when forced up over the treetops.

Wind incidence angle

To get the best efficacy, the windbreak must be placed perpendicular to the wind that you want to protect against. As the angle becomes greater or less than 90°, the amount of lee the windbreak provides gradually decreases.

Height

The height of the windbreak is one of the most crucial factors for the size of the lee side. How different heights affect the wind can be easily calculated even if the results are approximate because the other factors of the

REDUCED WIND SPEED

WIND

INCREASED WIND SPEED

WIND

The crown's shape and its effect

windbreak also affect your outcome.

On the windward side – the outer edge – the wind is affected 5–10 times the height of the windbreak. In an example with a windbreak of two meters, the wind is affected at a distance of 10–20 meters from the outer edge. On the lee side – the inner edge – the wind is affected 25–30 times the height of the protection, that is, in this case between 50–60 meters from the inner edge. The very best effect is found within the distance two times the height.

Density

The density of the planted windbreak mainly determines how significant the wind reduction will be on the lee side and how the lee is distributed. A sparse planting pattern gives a smaller area, but an evener lee over a longer distance, while a dense hedge gives a strong lee within a smaller area. Usually, the best solution is when the windbreak works more like a sieve than a firm barrier. The fact that the wind slows down and also warms up inside a windbreak is due to friction, caused by the wind hitting the plants.

A rule of thumb is that a windbreak is just right when you can see the colour of the ground on the other side, but not what grows there and that you can see movement on the other side, but not what is moving.

A tree avenue has a specific dampening effect on the wind over a slightly longer distance from the road itself; however straight underneath the trees and a little bit outwards the wind speed is instead increased, as it is

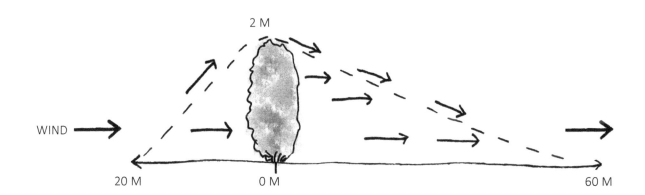

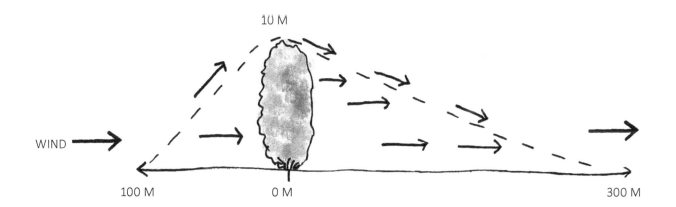

The height of the windbreak and its effect

pressed underneath the canopies. In some avenues, especially in old ones, one can see that the road surface is somewhat elevated in relation to the surroundings, which further strengthens the speed of the wind. The increasing wind speed over the road actually helps to remove the snow from it in the winter. Who knows, maybe this was one of the reasons why you used to lay long avenues over the open fields up to larger farms.

If an avenue, on the other hand, is to function as a windbreak directly next to the road, it must be supplemented with a bottom layer – for example, a hedge running parallel to the avenue itself. Since the majority of leaf plants shed their leaves in the winter, the difference between the effect of the windbreak in summer and in winter is quite large. One can expect that the leafless windbreak retains about 60 per cent of its protective effect. How much it really is is determined by how the plants' trunks and branches look. The more branched the plant, the better it protects, even without its leaves. An easy way to improve winter protection is to mix coniferous trees into the windbreak.

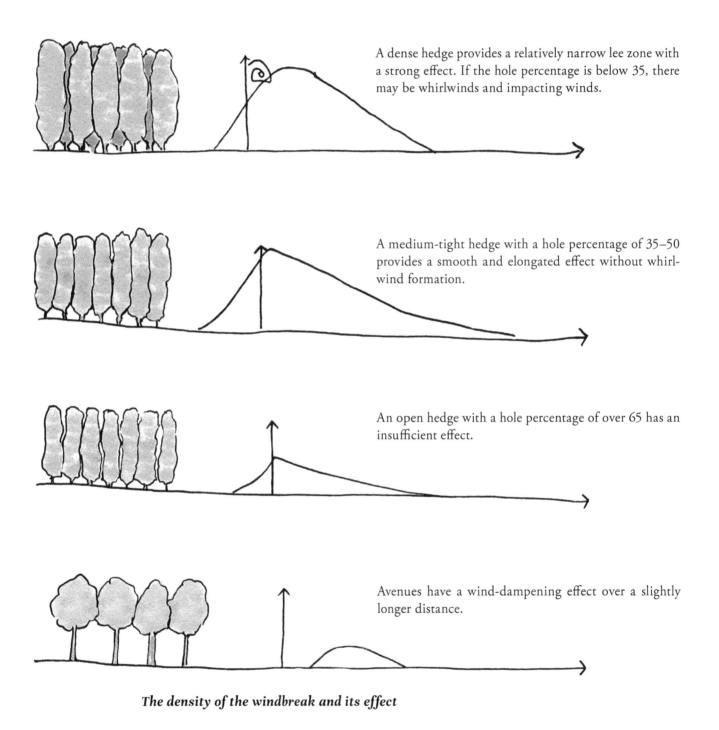

A dense hedge provides a relatively narrow lee zone with a strong effect. If the hole percentage is below 35, there may be whirlwinds and impacting winds.

A medium-tight hedge with a hole percentage of 35–50 provides a smooth and elongated effect without whirlwind formation.

An open hedge with a hole percentage of over 65 has an insufficient effect.

Avenues have a wind-dampening effect over a slightly longer distance.

The density of the windbreak and its effect

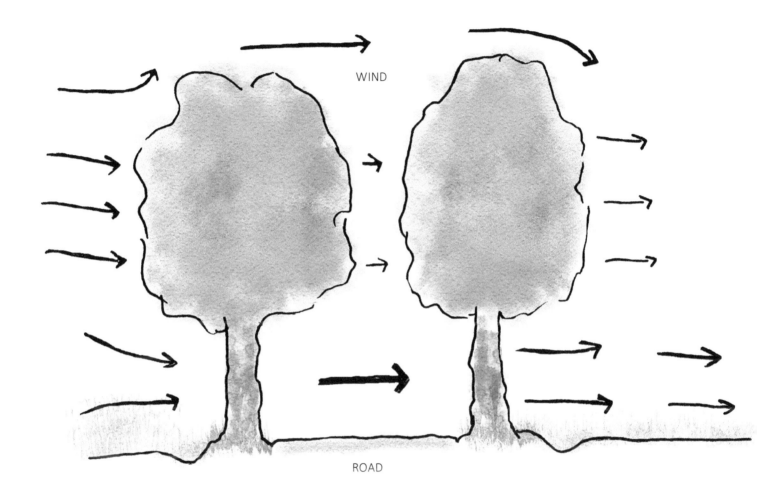

WIND

ROAD

The movement of the wind in an avenue

The shape of a windbreak

Windbreak width

The width affects the protective effect in relation to how trees and shrubs are placed. Trees in a single row provide almost no protection. When the row of trees is supplemented with a row of shrubbery, the protective effect is considerably increased. If you then make a pattern that consists of several rows, the protective effect can raise even more.

A wider windbreak that has dense shrubbery in the bottom and a more sparse upper layer gives the best effect. If you have problems with snow or soil accumulations along the outer edge, you can counter this by letting the edge itself be sparse, placing the shrubbery further in. Then both snow and soil will settle by the bushes inside of the windbreak.

Conversely, Too broad a planting pattern will have the opposite effect. The picture below shows how a very broad planting pattern without a layer of shrubbery results in a drastically lowered protective effect.

If you have a larger area that is too windy for crops or animals, you can plant several small windbreaks, evenly distributed over the area with a gap of about 20 times the height of the windbreaks. In agriculture, this is a great way to protect the crops on large open fields, but for this, a distance of 25–30 times the height of the mature windbreak is recommended.

Windbreak length

The length of the windbreak also plays its part as whirls are created around the corners of the windbreak. For a completely dense windbreak, this means that the ratio between length and height must be at least 12:1, that is, a one-meter tall windbreak must be at least 12 m long in order for the whirls not to affect the centre of the sheltered area. In permeable windbreaks, the ratio gradually decreases as the density decreases. At the same time, the sheltered area at the centre of the windbreak becomes larger.

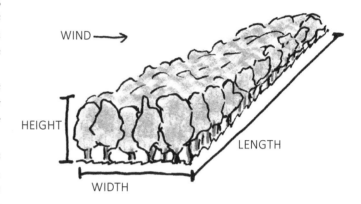

The shape of a windbreak

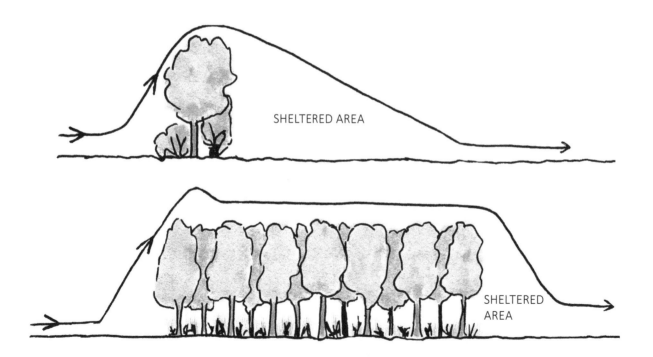

The length of a windbreak determines the size of the sheltered area

Plants as windbreaks around our houses

That which gives a house shade during the summer or winter season differs markedly.

A tree that stands relatively close protects against the high summer sun. In contrast, winter's low sun rays can shine through a leafless tree, even under the canopy itself. A deciduous tree near the house provides coolness in the summer and lets in light and warmth in the winter, but remember that even leafless trees create a certain amount of shade. A good rule of thumb is that trees with broad leaves give more shade in the summer, and less in the winter.

The best location of the tree is to the west, or at least to the southwest, which may come as a surprise. But when you leave the south side open, you get the full benefit from the winter sun. Even in the summer, you get the best effect, as the sun is at its hottest when it starts turning towards the west.

If you have problems with harsh winter winds from the north, it helps to plant spruces on the north side of the house. A well-placed curtain of trees can reduce the wind speed by up to 85 per cent, which reduces your energy consumption with 10 per cent, perhaps more depending on the house's own insulating ability.

If the house is in a very exposed location, one can advantageously plant a series of birches outside the spruces. Birch grows much faster than spruce, so the spruces are protected during their establishment. When the spruces have grown somewhat, it is advantageous to plant fruit trees and berry bushes on the inward side of them. This, of course, provided that the house does not cast a shadow there. When the spruce trees have grown large, you can gradually take down the birches, which will yield fantastic firewood.

To get the best effect from such a windbreak, it should lie 20–50 meters from the house. To obtain wind protection along with the entire house, the windbreak should be wider than the house with at least 20 meters in each direction. If you then finish the windbreak with deciduous trees at both ends, you will have something truly great, as these help reduce the turbulence around the corners.

Plants as windbreaks around our houses

A layout sketch for a natural windbreak

0 1 2 3 4 5

○ SLOWER GROWING TREES
✕ NURSE TREES
• BUSHES/SHRUBS

A natural windbreak and its structure
The best windbreak should be so homogeneous that it can be perceived as a single organic unit.

As seen from the direction of where the wind hits, the first row exists to stop the wind. The second row is for creating height and shelter in the upper part of the windbreak, and the third row for rounding out and stabilizing the inner edge of the windbreak as well as providing shelter along the base of the windbreak.

The most common design is one where you plant three to five rows, in which you systematically mix different types of plants, consisting of fast-growing trees – pioneer species, also called nurse trees –, slow-growing trees – late-successional species -, and various bushes.

The nurse trees grow rapidly gives the shelter height. In contrast, the more slow-growing late-successional trees mature, to eventually take over in the upper part of the windbreak. Arrow, for instance, is a pioneer tree species which quickly provides shelter. When the slow-growing trees have grown up, the fast-growing trees can either be completely removed or cut down to instead become a lower bush tree.

The number of fast-growing trees should be up to 1/3 of the total number of plants, the slow-growing trees between 1/6 and 2/6 of the plants, and the number of bushes should be between 1/3 and 2/3.

Below is a list of row- and plant spacing for hedge plants used in enclosures and windbreaks of various sorts. The distances are to be regarded as mean values since individual species have varying requirements for space.

The selection of plants
When choosing plants you should, as usual, choose plants of good quality and preferably local varieties. In order to increase diversity and at the same time reduce the risk of pests, several different species of one and the same variety are planted, both in terms of shrubs and trees.

When planning your windbreak, it is also important to know if the plant is a pioneer species that establishes itself quickly and efficiently, or if it is a late-successional species that is a little slower to grow.

A pioneer species is light-demanding and is easily outcompeted in shady environments. They spread efficiently through seeds or root sprouts, they hold their own against

Distance between plants when planting a natural windbreak

Plant spacing for deciduous trees and shrubs	In the row	Between the rows
Low, cut hedges	20–25 cm	
Free-growing hedges	50–100 cm	
Single-row windbreaks	100–150 cm	
Multi-row windbreaks	125–150 cm	125–150 cm
Shelterbelts with fast-growing trees	100–150 cm	150 cm
Final distances in shelterbelts	100–300 cm	300 cm

weeds, and they tolerate frost well. These types of trees are therefore good to use as starters in a windbreak, as they protect against wind, weeds - and together with bushes create a protective environment for the slower, late-successional tree species. If you are dealing with an environment that has extreme winds, it is wise to plant the fast-growing trees a few years before planting the rest.

In general, this system works well, but if the windbreak is in a less exposed area with good soil, the pioneer species can actually have the opposite effect and instead act as inhibitory. Then it is better to exclude them completely.

Late-successional species are, as previously explained, a little slower in the beginning, and they are relatively sensitive while establishing themselves. But at the same time, they live considerably longer. Trees like this are therefore ideal in a windbreak, for durabilities sake.

For a good result, you also need to take into account the type of environment a particular plant prefers as well as how tall and how wide it grows. More facts and a list of different plants can be found in the appendix "Plants for windbreaks".

A windbreak as seen from the side   *A windbreak as seen from the front*

Gardening for diversity and achieving an ideal harvest

How, and what we grow can be varied almost infinitely, as long as we understand what works best for our growing conditions and own goals.

The ideal place to grow your crops

The very best place for cultivation here in the Nordic region is on a southern slope; due to this being the angle where you get the most out of the sun. Most other conditions can be solved when the place is right.

The location should have protection both to the north and to the west. The soil needs to have rich humus content, be nutritious, and be able to retain a fair amount of water.

The cultivation beds should first of all lie slightly twisted to the southwest, as long as this does not mean that they end up sitting vertically in relation to the slope. Should this be the case, it is better to lay the beds horizontally along the slope, and adapt how you plant in the beds so that the crops face the southwest.

An old piece of Swedish wisdom says;

"The north wind brings frost and hail, the chilly westerly winds delay the development of the plants, and may even kill them".

— From Trädgårdsmästaren, in 1917

Choosing plants

Among the most enjoyable things to do is reading seed and plant catalogues – but sometimes it is important to master oneself, and plan for what you actually need. Here are three important factors to consider:

→ *The shape of the plant, how it grows and how old it becomes*

→ *The plant's climate zone, how it tolerates sun, shade, what it needs from the soil, pH tolerance, and other specific requirements that a plant may have*

→ *Use of the plant; such as food for humans and animals, to attract pollinators, to prevent damage, medical use, soil improvement, soil erosion protection, wind protection or simply just because it is beautiful*

Create diversity

Diversity is invaluable in many aspects. When it comes to cultivation, it is easy to focus only on what benefits yourself. Still, to create a good ecosystem, other plants are needed. Here are another three key points that one needs to incorporate to create a greater diversity:

→ *Host plants for insects feeding on pests (predators)*

→ *Sacrificial plants are plants that are sacrificed to various pests to protect the plant that you want to prioritize*

→ *Plants that attract different useful insects and pollinators*

Improve the Earth

The most important thing when cultivating is actually "feeding" the soil. If you have poor soil, it is natural to improve the soil to grow plants. Still, if you do not return the nutrition that disappears when you grow and harvest, the soil will gradually deteriorate and eventually die. Just like all other living things, the soil also needs food.

Grow vertically

Many plants prefer to grow up along nets, trellises, poles, fences, and trees. You can even use hanging flowerpots of varying kinds. Plants that grow vertically have the great advantage of only taking up a minimal surface area to its size. These plants also protect the soil around them. This can be used advantageously in your garden by, for example, planting a vertical plant next to another low growing plant that needs some extra protection. The more powerful climbing plants can also be used to create a shaded trellis at the patio, or on a hot south wall to cool the house.

Intercropping

Intercropping, or co-cultivation, has become increasingly common, which is a huge advantage for everyone involved; the soil, plants, insects, and ourselves. Intercropping is a subject where there is a lot to learn, but the foundation is that you match up plants that support each other in different ways. It can be plants that do not compete for the same nutrients or space (both above and below the ground) or a sun-loving plant that gives shade to a sun-sensitive variety. Another form of co-cultivation is to combine perennials with annuals. Because the perennials are usually the ones that get going first in spring, the growing season is extended, and the yield from the same area is thus increased. At the same time, the plant waste from the one-year-old plants nourishes the perennials if this waste is allowed to remain and decay.

Many plant combinations provide mutual support for proper growth. A classic example is the one from Native Americans, called "the three sisters". It is a co-cultivation of maize, beans, and squash where all three are better off when planted together. The maize plants become climbing support for the beans, which turn into a windbreak for the squash that grows along the ground. The squash, in turn, helps the others by keeping the weeds away. In addition, the beans also provide extra nutrition to the soil.

Intercropping examples

Here are a few of the many different co-cultivation combinations available. The appendix contains a more detailed list of which plants get along with others and so on, which can be useful to have at hand when planning your crops.

Peas provide nitrogen to the soil and are the tallest. Therefore, they should be at the back. Carrot and onion protect each other against different pests. Tagetes protects against the leek moth and is tasty in salads while adding a sense of beauty to your garden. It is preferably placed between the onion and the carrot; both of which are narrow in their growth pattern.

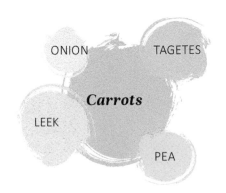

Jerusalem artichoke protects tomatoes from cold winds. Tagetes protects against nematodes and again, is good in salads. The tagetes are preferably planted in front of the tomatoes. Borage can help achieve an earlier tomato harvest. It is, however, essential to keep it in check, due to it being so invasive. If you still want to try it, it should be planted in the outer edges of the tomato area or in pots.

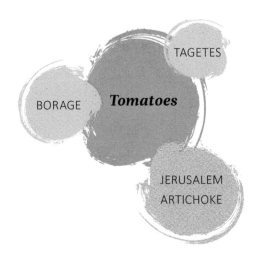

Chervil protects lettuce against aphids, but keep in mind that chervil is an enthusiastic seeder and can cause some troubles the following year. Since leeks and beets grow slower than lettuce, they do not compete for the place and light that the lettuce needs to grow well. Once the lettuce is harvested, both leeks and beets have room to grow large.

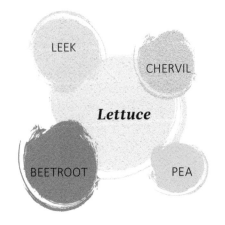

Wormwood protects against pests due to its strong fragrance. Both wormwood and chives also protect against Puccinia ribis, "Currant rust". Fragaria vesca (wild strawberry) and Arctic bramble are good ground covers while at the same time providing tasty berries.

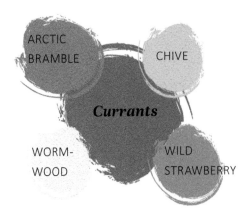

Beans provide nitrogen to pumpkins. Onions and spinach have shallower roots and therefore do not compete with the pumpkins deeper roots. Black salsify is grown between onions and spinach because it benefits both plants. Finally, lacy phacelia for green manure. Think of the heights and place the beans at the back and the spinach at the front.

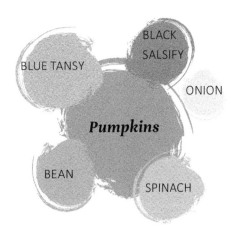

Roses, together with all of these herbs, create stunningly beautiful beds. The chives protect the roses against black spots and enhance the scent of the roses. Hyssop and thyme protect against aphids.

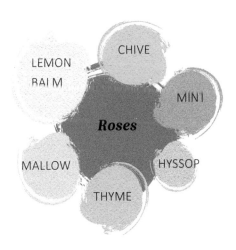

Natural group of plants

Cluster gardening

In most gardens, apple trees stand alone on the lawn. Nothing grows beneath them, and the surface beneath them is a bother, where the grass needs to be cut and fallen apples removed. But it does not have to be like this!

Many plants thrive in the walking shade along the outer part of the tree crown. Why there are even plants that make the apple tree healthier and provide a better harvest. With a broad diversity in plants, the variety of animals will also increase. Another important detail is that when the rain passes through the tree, it brings with it lots of different nutrients on its journey down to the ground. This rain can contain as much as 50 times more nutrients than regular rain. This, of course, benefits the tree and everything that grows underneath it.

Growing in this way is usually called cluster gardening. It is an effective form of intercropping based on the creation of interaction between plants, animals, and other elements in nature. Often, the clusters are grouped around a central element. On larger surfaces, several clusters can be connected.

Let us take the apple tree as an example of a central element where the following parts interact around to create a well-functioning intercropping.

→ *Comfrey as green fertilizer*

→ *Gagea lutea as ground cover before the apple tree has its leaves and other perennials have sprouted*

→ *Plants that deter pests*

→ *Plants that attract animals that feed on pests*

→ *Plants that attract pollinators*

→ *Legumes that bind extra nitrogen to the soil*

→ *Nitrogen-fixing trees – alder, for example*

→ *Rosehips as protection for visiting birds*

→ *Trees and shrubs that can protect against wind, sun, and frost*

→ *Even animals fit well in such a cluster, either for shorter or longer periods*

→ *Chickens for a shorter period to shift, scratch, and fertilize the soil*

→ *Ducks who eat snails and slugs*

→ *Guard dog to keep wild pests away*

Getting all these parts in a regular garden can be difficult. Still, selected parts will get you a long way, and the ground underneath the apple tree finally has a use.

Imitate wild nature

To increase your understanding of the benefits of intercropping in clusters, one can look at the wild nature. There are plants at all levels; tall trees, low trees, shrubs, creeping, and winding plants. But only things that thrive in that particular place. To the untrained eye, this can look messy, but if you look closer, you can see that nature fills in all the "voids" that occur in the environment. No bare ground or "empty holes" in the air are left to chance. All plants fulfil their function when nature itself decides. This can be advantageously mimicked in our gardens by thinking this way.

As tall trees, large and medium-sized fruit trees are suitable. Low trees can be different forms of smaller fruit and nut trees. At the shrub level, there is a lot to choose from with varying bushes of berry: next level down, different herbs and perennial vegetables. And finally, on the ground level, it is advisable to use some form of creeping ground cover. If wanted, you can add plants where the roots are the vital part, and preferably some climbing plant along any tree trunk or trellis. Incorporate all this, and it begins to resemble wild nature.

In the garden

Back to that little house garden. Here I have chosen to illustrate a more straightforward intercropping cluster that can fit in many gardens.

Under the apple tree, there is now instead of just grass lots of edible plants. Near the tree trunk grows hosta, whose crispy sprouts in spring are good to eat. Further on in the season, they will bloom beautifully. Here you can also plant gagea lutea or scilla, which bloom before the apple tree has set its leaves. You can, of course, choose other early, bulbous spring flowers – you can also spread them out in the white clover area outside. This area fills several functions. Partly, it provides easier access for picking apples and other plants. Secondly, white clover clippings are excellent fertilizer that you can advantageously spread out around the plants. If you have dandelions, please let them stay for the benefit of the apple tree.

In the entrances between the bushes and on the sitting area there are stepping stones and in between, Thymus pseudolanuginosus.

In the east, there are two different kinds of honey-

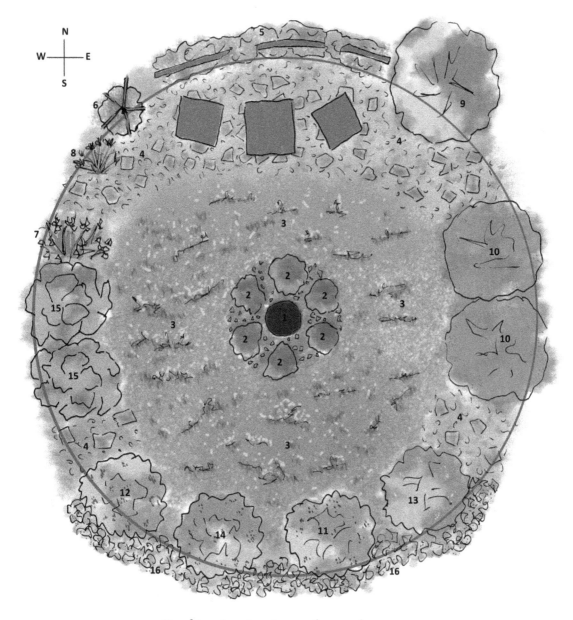

Combination growing underneath a tree

Plant list
1 TREE TRUNK
2 HOSTA
3 WHITE CLOVER
4 THYMUS PSEUDOLANUGINOSUS
5 BLACKBERRIES
6 HABLITZIA
7 TREE ONIONS
8 HYSSOP
9 SILVERBERRY
10 HONEYBERRIES (2 TYPES)
11 BLACKCURRANT
12 REDCURRANT
13 YELLOW GOOSEBERRY
14 RED GOOSEBERRY
15 RHUBARBS
16 FRAGARIA VESCA
(WILD STRAWBERRY)

Other
Stepping stones in the
walkways and for
the sitting area
2 chairs and a table
Trellis for blackberries
and Hablitzia

The gray circle marks
the approximate size of
the tree crown.

berries, in the south, currant and gooseberry bushes with Fragaria vesca (wild strawberries) in front. Rhubarb grows in the west and next to it an area of tree onions. Onions of all varieties can help keep some pests away.

Behind the sitting area on the north is a trellis with blackberries that not only provides tasty berries but also acts as a windshield. To the right of the blackberries stands a silverberry bush; as a windshield, to help to give an increased harvest, and because it smells so good. To the left of the blackberries stands a small trellis with Hablitzia. Next to the Hablitzia grows the herb hyssop which is especially loved by bumblebees.

Perennials for increased durability

Perennial vegetables are still used far too little, but a change is on the way. The advantages of perennials are that they start earlier in spring and they are usually more climate-friendly thanks to their more extensive root systems. Perennials thus provide a more sustainable cultivation system compared to if you only grow annual plants. To make the system especially sustainable, it is usually best to have several different varieties; Some that can withstand a cold and wet summer, some that can withstand a hot and dry summer, and as a base; some that thrive best during an ordinary summer.

Extend the season

Since a plant only grows during the part of the year that the temperature stays above five degrees, we up here in the Nordic region benefit massively from helping nature a bit on the way to stretch the growing season. You can do this by working with microclimate in different ways, growing in a greenhouse or perhaps even in a winter garden if you have the opportunity. Other methods are to grow in both cold and hotbeds, to grow in raised beds, to mulch, and to cover with cover fabrics or similar materials—all to cheat both the last and the first frost. With a little inventiveness, you can extend the season with a justifiable number of weeks.

Start harvesting early

Many of our plants can be harvested earlier than traditionally. A good way is to plant too tight or cull too little. Then, you can harvest baby versions of most of our common root vegetables. Many of the plants' leaves are also excellent in a salad, such as spinach and beetroot.

Indigenous and local plants

By choosing native, and preferably local varieties, one makes it easier for themselves – because the local plants have developed over many years, (centuries even!) and they are thus adapted to the environment and the climate of that particular place. This applies not only to all plants but also to all animals. When it comes to so-called domestic animals such as chickens, ducks, sheep, etc., several country breeds originate from different local parts of the country. Helping to preserve these is very important.

Winter protection

By putting some effort into helping the herbaceous plants prepare for their winter dormancy, the durability of the crops is increased. Four good ways are:

→ *Cover the soil with leaves, straw, spruce twigs, or perhaps even with an old carpet. This way, the soil is isolated, and the risk of frost damage is reduced. Meanwhile, the microbial life reaps the benefits, since their work season is extended in the autumn*

→ *Work on keeping the soil well-drained. Sandy soil is in itself well-drained and deals with the drainage on its own*

→ *If you have clay soil, it is best to grow in raised beds where you add large quantities of organic material*

→ *Last but not least, one can work on creating a microclimate around the plants*

Edges in practice

Nature becomes the most vivid where different environments meet, and we call these border areas edges.

The fact that the majority of people live in edges – especially in-between land and water of different forms – is not a coincidence, but rather a natural consequence of the fact that the highest biodiversity is found here – and thus the greatest access to food.

An example of an edge that is easy to observe is the forest edge. When trying to enter a natural forest, the edge effect is clearly marked out—densely populated with everything from small bushes to large trees, with a lot of birds and insects. Once you have entered the forest, it opens up, and it gets quieter – because the diversity here has decreased.

Animals are also looking for different edges due to the diversity in them. One example is deer, who likes to stay around the forest edge or along the edges of the fields. Even fishermen prefer to go along edges of different kinds – like the edge of the reeds, or in the shade of a jetty – instead of fishing in the middle of open water.

Edges of different shapes

The edge effect can be utilized in many of the garden's various areas, and its multiple forms.

Ponds

In nature, as mentioned, the edge between land and water is the most fertile. A pond in the area provides the opportunity to recreate this rich environment. But making a round pond is not the most efficient, nor a square for that part; creating a pond with the longest shoreline possible is. You do this best by letting the shoreline be irregular and wavy to create the maximum space for different plants and animals. If you then add a small island, the water depth varies. If you then add a rough tree branch or an old log, you have created perfect conditions for a place full of life.

Beds

The same idea applies when planning different garden beds. A straight line gives the smallest edge effect, so a wavy line is much better. A few examples of effective forms are the keyhole shape and the spiral shape.

Planting

Again, same when you sow and plant. There are fewer plants in a straight line than in a wavy one. In addition, a wavy planting pattern means that each plant has a bigger border against its surrounding environment with, for example, more sun.

Vertical gardening

Vertical gardening also increases the edge effect on trellises and against walls and fences; yet another edge is created – the height. The height means that more insects can be found there, that a windbreak is created, and that it can provide shade and protection for what grows on the ground.

Windbreaks and demarcations

Again, the wavy lines provide the best benefit. Just look at the natural forest edge, there are no straight lines – because when nature takes care of itself, the line swells back and forth. Straight lines in nature are human inventions.

Examples of edges

The edges along animal fences are valuable areas that you can both benefit from and reinforce. Because animals like to stay close to the fence, a lot of manure is collected there. Not only does it fertilize the soil there, but a lot of the nutrition also flows into the adjacent water, such as a ditch. If you can take advantage of this nutritious water and lead it on to your crops, much is gained. Another way to gain a greater benefit from an animal fence is to plant trees along the fence – for both food and shade for the animals.

Another example that can give more trouble than joy is to put stones around flower beds and crops. Because around these pretty stones, weeds will soon sprout, and then you will most likely go about muttering over the issue. The stones were only put there to keep the grass away in the first place. The reason? The edge effect explains everything. The stones work as a water collection system in miniature, as well as having some help from the wind to blow in nutrients between the stones. The result is a perfect microclimate where plants thrive. And then we wonder why it got to be so bothersome!

The problem can be solved by growing something on the inside next to the stones, which then reaps the benefits from the stones' ability to collect nutrients and water. The outside is a little trickier, at least if you want a walkway there. If that is the case, it should be raised so that it is at the same height as the stones, with nary a space between the different surfaces. This is not easy to arrange. It is easier to omit the rocks and, instead of a path between stones, perhaps make it out of stepping stones with, for example, Thymus pseudolanuginosus between the stones. Beautiful to look at; smells good when you walk, and you don't have to clear weeds.

Another example is how leaves, snow, and even debris gather against a fence. This can be a negative if you do not think about it when planning, and it can be woven into the planning to counteract snowdrifts where you do not want them.

If you do not know how to plan different surfaces and edges, the trick is that you start by planning for what SHOULD be along the different edges. Plan how the boundaries should be managed and what should be grown for the best possible results, then the rest will resolve itself. Even though it may sound strange, the surfaces (the spaces in between) will come along with the bargain.

Once you have learned to see the edges, you will find that they are astoundingly many. Here are a few examples of an infinite variety of variants:

→ *Field/Forest*

→ *Lake/beach*

→ *Farmland/wild nature*

→ *Wall/house drainage*

→ *Wall/cultivation area*

→ *Walkway/lawn*

→ *Fence/lawn*

→ *Driveway/lawn*

- → *Culture bed / logs*
- → *Logs/lawn*
- → *Culture bed/stone wall*
- → *Stonewall/lawn*
- → *Culture bed/fence*
- → *Pond/cultivation area*
- → *Growth area/lawn*
- → *Pond/lawn*
- → *Culture area/boulders*

- → *Sun/shade*
- → *Acid soil/neutral soil*
- → *Roses/chives*
- → *Carrots/leek*
- → *Cabbage/celery*
- → *Lettuce/chervil*
- → *Strawberries/beans*
- → *Squash/nasturtium*
- → *Raspberry/spinach*

EDGE ZONE

EDGE ZONE

Edge zones between pathways and garden beds

Cultivation Beds

A cultivation bed can be at ground height or more or less raised. It can be drawers and pots on a small balcony, and it can be many large beds on a field. The raised cultivation beds have many advantages both from plant and work point of view compared to those that lie directly on the ground. The height means that they have a warmer climate, and if they are also arched, they create a larger cultivation area. A fascinating figure is that a well-planned balcony of 20 square meters can provide as much as 70 kilos of food in one year.

Plan cultivation beds

First and foremost, it is the size of the place that determines how much you can grow. Once you start planning, you should do so based on what is needed. Growing for the sake of growing is rarely a good idea. Instead, it can be counterproductive when you can't manage it. Better to start on a smaller scale and experience the joy and even the craving for more crops the following year. A smart tip is to start with something that you like to eat.

At this point in the planning comes the next step. How large beds do I need, where should they be, and should they be raised? Plants to be harvested every day or quite often should be located closest to the house – that is, in zone 1 – while plants harvested only once or a few times a week may be located further away from the house. Remember to place the beds to get the most sun. The very best is if you can direct the vegetation more to the southwest than straight to the south – because then you also get the evening sun's last warming rays on your crops. If it is not possible to have the beds in that direction, at least try to make sure that the plant rows go in that direction, and that the lowest plants are closest to the sun.

Some places on the beds are more vulnerable than others. The most sensitive points are the corners where the aisles or other surrounding environment dominate. To create stability and secure the corners, it is good to plant perennials there. The next sensitive area is the short sides of the beds, where it is also good to plant several lower perennials, such as herbs. If you have no built-up sides around the raised beds, the lower edges of these beds are also vulnerable. Here too, it is suitable to grow herbs or preferably strawberries.

The next step is to go through and plan which plants to grow. Use the technique to grow with crop rotation and find out which plants thrive best together. Also, keep in mind that the highest plants should be placed in the middle of the bed and the lowest along the edges – everything to be able to harvest easily and comfortably.

Material selection in the beds

Remember to avoid using materials that result in different chemicals ending up in the beds. Here are some examples of what to avoid:

→ *Treated wood*

→ *Straw where the grain has been sprayed with various pesticides. This can destroy the whole cultivation area for many years to come*

→ *Fertilizer from stables using a lot of antibiotics and pesticides*

Raised beds

Raised beds are the most straightforward shortcut to a better growing climate. Building raised cultivation beds is also a quick way to get better drainage. In order to gain a plant zone, it is usually sufficient to apply more soil so that the cultivation area is 15–20 cm higher than the surrounding area. The effect is especially noticeable in the spring when the soil in the raised bed is heated faster. With a little luck, you can start your gardening several weeks earlier than what you ordinarily would – but remember that the sooner you start, the higher the risk of frost. Garden fabric is an excellent solution to the problem, and you can readily have it on 24/7 for the first few weeks while the plants are establishing themselves.

The height of the bed can be achieved in several different ways, where the lowest bed, as previously stated, only has a lot more soil in relation to the walkways. The next variant is that you make a base with straw, manure, old leaves, et cetera, that you then put the soil on top. The next variant is where you start with branches and twigs for a base. Finally, you have the variant where the base consists of logs and thick branches. In order to avoid soil compaction in the finished beds, you must never walk on them. Later, I will take you through the construction of the higher beds.

Of course, the garden beds don't need to be tall, but it tends to make gardening easier. One significant advantage of raised beds is that they are considerably more productive because they are initially made up of lots of compostable material. By continuing to feed them with the more nutritious matter, they retain their fertility.

Raised bed with hügelkultur

Hügelkultur is a German word, meaning hill/mound culture. It is a unique form of tall beds that are becoming more common, especially in permaculture. The idea is not new; it is an old technique that has been used for centuries in both Eastern Europe and Germany. Interestingly, the hügelkultur technique must go through cold winters to function, which is very suitable here up in the north.

The benefits of the hügelkultur

The benefits of growing in hügelkultur beds are many:

→ *No digging needed*

→ *Less work and better working posture*

→ *Much of the kitchen and garden waste is used*

→ *The waste is composted directly on site*

→ *You do not have to buy expensive soil or spend a long time creating a good compost soil*

→ *The bed maintains an even moisture level and reduces the need for irrigation*

→ *The bed creates a nutritious and viable soil*

→ *Combined with mulching, almost no weeding is needed*

In literature, there are many "recipes" on how to create the perfect bedding. Still, the basic technique is simple: Build the beds of various organic compostable materials, place the coarsest in the bottom, and fine soil on top. Then, simply let nature have its course.

The difference between hügelbeds and other forms of beds

There are some important differences between the hügelbed and other forms of beds – differences that explain the benefits of hügelkultur.

Wood base

Hügelbeds are built from a base of logs, branches, and twigs from woody shrubs. The more wood and the coarser the wooden material is, the more self-sufficient the beds become over time. You can even use old stumps and roots.

Framed Hügelkultur on flat ground

The size

Hügelbeds can be huge, and some beds are over two meters tall with whole trees as the base material. These large beds not only provide a large cultivation area, but they also function as windbreaks. At the same time, in principle, no irrigation is needed.

But even if the large bed has its advantages, 60–90 cm is a more suitable height in a regular garden.

The slope of the sides

Most other bedding methods create relatively flat beds, but the hügelbed expert Sepp Holzer disagrees that this is ideal. Instead, he advocates a 45-degree slope so that the beds will not be too compressed since an over-pressed bed will result in lower oxygen supply to soil and plants. But even a flat hügelbed can work well if you take care to add more organic material continually.

Nutritionally superior

Because wood rots slowly, this subsequent degradation provides nutrients to soil and plants for a considerably longer time than the beds made without wood.

A large bed can provide a constant supply of nutrients for 15–20 years (or even longer if only using deciduous trees). The composting of wood also generates extra heat which prolongs the growing season, especially in the first few years.

Retains moisture

When wood decays, large amounts of fungal mycelium are formed. Not only does this provide better nutritional conditions; it also creates countless small air pockets that help keep a stable moisture content in the beds. Provided that one-year-old plants are not cultivated with shallow roots, the smaller beds can survive for several weeks without irrigation, while the largest can cope with whole summers without extra irrigation even in dry climates.

Planning your hügelbed

Since a hügelbed lasts for many years, it is essential to plan appropriately before deciding where it should be and how it should look.

Just like in almost all forms of cultivation, it is essential to get the most significant benefit from the sun. The right direction, that is, southwest, is especially important when it comes to hügelbeds as the height and the inclined sides may create a permanent shaded side if you put the bed in the wrong direction. If you have problems with wind, it may be worth considering the wind direction when planning to see if the bed can also function as a windbreak.

Even the slope of the ground and the natural water flow are essential factors that one has to take into account. Because while a strategically placed bed can help to suck up excess water, too much water can cause problems and begin to undermine the bed. Should there be problems with excessively high water pressure or muddy soil, it can be solved by adding a layer of large stones as drainage underneath the wooden layer.

What the bed looks like is only determined by the imagination. It does not have to be straight or square, but it should preferably be rounded, have wavy edges, be horseshoe-shaped, or why not a little winding.

Depending on the size of the project, one may have to start collecting material the year before to get enough together. Logs, branches, twigs, and roots are the ideal base for a hügelbed.

Untreated wood is good to use as well as wood chips, but keep in mind that the more finely divided the wood is from the beginning, the shorter the life of the bed. Also, it quickly becomes too compact.

If you do not have direct access to wood, you have to be creative and ask the neighbours – or even the municipality you live in.

The function of the wood and various types of it

It is the decaying wood that is the magic behind this technique. Of course, you can start with fresh wood, as long as you take into account that the yield will not be the best during your first year.

The harder the wood, the slower the decay – and the longer the bed will remain productive. Coniferous trees are okay to use, but only to a lesser extent because they provide a more acidic soil. At the same time, they rot a little too quickly. If, on the other hand, you want to grow plants that want acid soil, wood from coniferous plants is instead preferable.

The best tree varieties are apple, aspen, birch, poplar, maple, oak, willow (make sure it is dead; otherwise, it will start to grow).

However, some types of wood should be avoided, and these are cedar, Robinia (false acacia), black cherry, black walnut, and treated wood. They emit substances that are unfavourable to the cultivated plants in various ways.

Building a hügelbed

Once you have decided the place, how the bed should look, and gathered all the material, it is time to start building.

Depending on how the ground looks where you plan to build, you can start in two different ways:

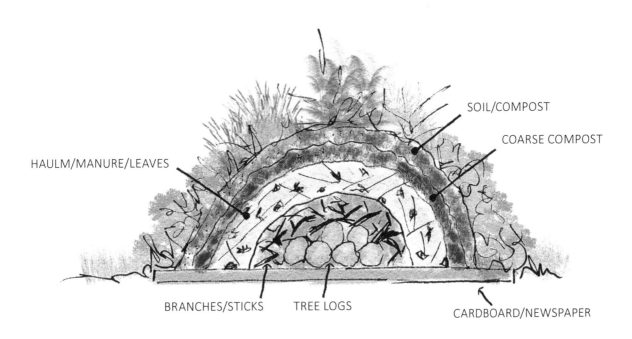

SOIL/COMPOST

COARSE COMPOST

HAULM/MANURE/LEAVES

BRANCHES/STICKS

TREE LOGS

CARDBOARD/NEWSPAPER

Intersection of an Hügelkultur

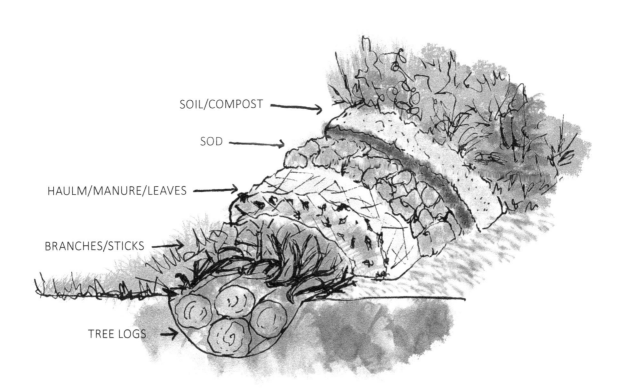

SOIL/COMPOST

SOD

HAULM/MANURE/LEAVES

BRANCHES/STICKS

TREE LOGS

Perspective of an Hügelkultur

→ Either dig a 30–50 cm deep trench and lay the turf rolls and soil to the side for use at a later stage

→ Or if there are aggressive weeds or if the soil is otherwise inappropriate, you build the bed directly on top of the ground with a thick layer of cardboard as the first layer

→ If you have planned a flat bed with a frame around, it is now time to make the frame.

→ It is essential to soak each layer before the next one is added

→ Place out all the wood. Start with the larger and coarser pieces in the bottom and in the middle. If you have a truly muddy soil, it is better to use short logs that you set up vertically in the dug ditch

→ Then spread out the smaller and thinner ones on top of the coarser. Try placing and pushing together all the wood so that it is steady before you put on the next layer

→ The next layer consists of finer twigs. If you are spry, you can now step up onto the bed; otherwise, you can push it together to the best of your ability by hand. This is done so there will not be big air pockets in the bottom, as these pockets mean that the bed deflates a lot during the first few years

→ The next layer may consist of a little everything; tufts of grass (with the turf down), straw, old hay, silage, leaves, kitchen compost, or semi-finished compost. Most organic materials work well.

→ Depending on what you have, it is also advisable to mix well-aged horse or chicken manure preferably with straw or bedding. When you start laying the material, you need to make sure that it also reaches down among the branches and twigs, without pushing it down. How thick the layer should be you decide yourself, but estimate at least 60 cm because it compacts very quickly

→ Finally, you put on a layer of finished compost or soil with anything from 30 cm to just a few centimetres. How much is determined by what you intend to grow the first year

You get the best results if you build the bed in the fall so that it can mature a little before it is time to sow. But of course, it is possible to start cultivating pretty much right away, if you keep track of the temperature so that the composting process does not make it too hot in the beginning.

If the beds are on a slight slope, it is good to make it easier for the beds to suck up surrounding groundwater by digging the aisles about 10 cm deeper than the surroundings. Then you cover the aisles with, for example, mulch.

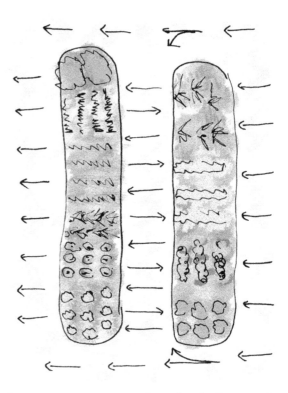

Water pressure around garden beds on a slope

Hügelkultur on a slope

If there is a lot of water that presses up against the beds, it is preferable instead to do the swales that I described earlier.

Lasagna bed

If you do not have access to tree trunks and twigs, the so-called lasagna bed is an excellent alternative to building a raised bed.

This shape is also suitable for smaller beds, for example, when growing in pallet collars. The lasagna bed is also built up of different layers. Remember to water each layer properly before adding the next one.

→ *Finely divided green forage*

→ *Soil improvements such as blood meal, stone meal, and seashells*

→ *Fertilizer, a thin layer of, for example, chicken manure*

→ *Newspapers or cardboard. At least two layers of newspapers, but preferably up to 2.5 cm thick*

→ *Well-aged manure from, e.g. horse or cow. 1–2.5 cm thick (without straw)*

→ *Hay. 20–30 cm thick. Alternatively, if you cannot get clean fertilizer, you can replace the two above layers with a layer of straw-mixed fertilizer. Remember, however, that straw takes nutrients, while hay adds it*

→ *Compost. If you are going to cultivate immediately, the compost needs to be finished. If you make the bed in the autumn to grow the coming spring, unfinished compost works well.*

→ *If you want to get a really fine-grained top layer, you can top it off with a thin layer of planting soil.*

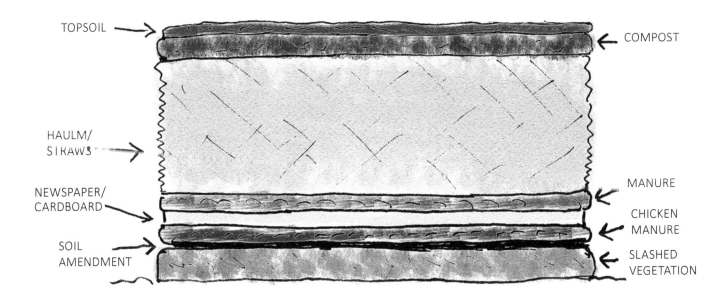

The different layers of a lasagna bed

Keyhole beds

Another space-saving cultivation method is the so-called keyhole bed, which in some forms is called a "Mandala". Planting in circles is also more effective than growing in straight lines. In a keyhole bed, you get a higher return per square meter.

The keyhole bed can either be a freestanding circle or combined into a series with several keyholes in a branch from the main path. The central path may well have a winding form as it increases the number of meters of easily accessible edges. How to choose your keyhole structure depends entirely on how much space you have available, as well as your mobility and the need for space. If space for wheelchairs is needed, the actual keyhole space is made slightly larger, while the height of the beds is adapted to suit a wheelchair. A variant of a smaller keyhole bed is one where you replace the hole in the middle with a compost, which provides the bed with nutrition.

The description that follows explains how to make a raised keyhole bed consisting of only one circle. Remember to soak each layer during its construction.

→ *Start by marking out a circle where the bed will be. Size and height are determined by several different things; space, access to materials, what to grow, and potential disabilities that you want to take into account. A good benchmark is a diameter of about two meters for a circle*

→ *Then mark out a smaller inner circle in the middle of the larger circle. If the bed is two meters in diameter, it is appropriate that the centre circle is 30–40 cm in diameter. Then, mark out a path, either straight or wedge-shaped, between the inner and outer rings. The width of the passage is again determined by the space requirement you have. Even if you do not need to accommodate a*

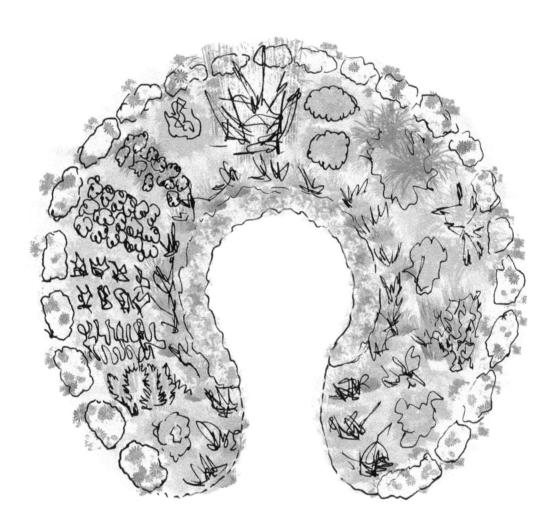

A simple keyhole garden

wheelchair, you may want to be able to pass with a wheelbarrow

→ *If you want to build tall beds, walls are needed to keep the bed material in place. Start by creating the outer wall. Most things will work; bricks, stones, wood, or metal. If the bed is going to be lower, logs also work as outer edges. Keep in mind that different materials create different microclimates where, for example, the stone retains more heat than what wood does. When the exterior walls are done, you continue with the rest of the walls*

→ *Then fill the cultivation bed as desired*

→ *If you do not have access to wall material or if you want a lower bed, you can still make a keyhole bed. Mix brown and green composting material in the ratio of three to one. The brown can consist of old leaves, straw, sawdust, newspapers, cardboard, and branches. The green can in addition to green plant material also include kitchen waste, composted manure, and grass clippings*

→ *As usual, the coarse material is laid in the bottom, finer in the middle, and soil on the top. The thickness of the different layers depends on how high you want the bed, and of course on what material is available. Keep in mind that the bed deflates quite a bit during the first year*

Finally some planting tips specifically for keyhole beds. The fast-growing plants that are harvested are often placed closest to the centre, such as lettuce. Next, the plants that take a little longer, and outermost those that take the longest time, such as potatoes and carrots. If there is any plant that needs extra wind protection, you can grow something taller next to it, on the side where the wind is harshest.

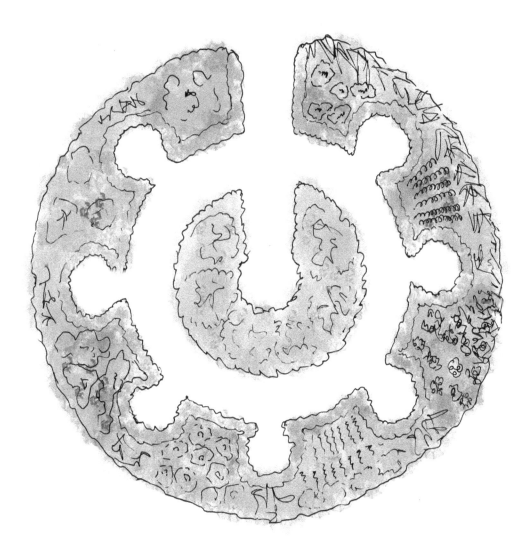

An advanced keyhole garden

Herb spiral

Another functional and often used form is the so-called herb spiral.

Suppose you imagine a long bed with plants, which you then grab and wrap around yourself. Then; you have created a spiral where the starting point in the middle should also be the tallest, and then fall in step with the spiral turning counterclockwise around the centre point to come to a stop at ground level. Where the bed ends, you may have an itty bitty pond with water for insects and small animals.

This way offers a lot of room for cultivation on a tiny surface compared to a single, long bed. The size of the herb spiral is determined by how far you can reach with your hand. It usually means a diameter of about 160 cm and a height of about 1 meter. The width of the cultivation surfaces should be around 30 cm. All in all, the herb spiral provides a varied cultivation climate from dry to wet, where it is also sufficient to only water the top.

The benefits of a herb spiral

The advantages of a herb spiral are many:

→ *Aesthetically beautiful*

→ *More heat thanks to the stones*

→ *Smaller weeds, thanks to a smaller edge effect towards the surrounding environment*

→ *Harvesting the herbs is easier*

→ *Good working height*

→ *More cultivation on a smaller surface*

→ *Effective water use*

→ *Provides different microclimates with warm, sunny, and dry at the top of the south side – and more cooling and more humid at the bottom of the north side*

An herbspiral as seen from above

At the top of the spiral, you grow the most sun-demanding and drought-resistant plants such as thyme, lavender, sage, and rosemary. The farther down you get, the more moisture-consuming plants thrive. At the bottom, plants such as parsley, coriander and various sorts of mint are suitable.

When choosing your herbs, it is good to avoid those that spread a little too well, for example, mint and coriander. Depending on which herbs you choose, you can fit 30 different herbs in such a herb spiral.

When comparing the edges of a long narrow bed and the outermost border of a herb spiral, the difference is significant. Along the long narrow bed, a long and troublesome edge arises with the risk of a lot of weeds. On the other hand, in the herb spiral, all edges are used except the outer part of the lower circle.

A beautiful and practical arrangement is that you incorporate a water hose inside the bed up to the top, where the hose ends with a small sprinkler. When you then water, you need less water while the irrigation itself becomes a feature to look at.

Building a herb spiral

→ *Start by deciding whether it needs digging, or if the ground is to be covered with cardboard and newspapers*

→ *Then, draw out the outermost circle and then mark the inward spiral. If you start from the centre, the spiral is drawn counterclockwise, and if you start from the outside, it is drawn clockwise. Count on 30 cm cultivation area between each turn.*

→ *When you are satisfied with the shape, it is time to place the edges out. Different materials such as natural stones, bricks, or logs of wood are set up in a row after each other.*

→ *If you have access to sheep's wool, it is excellent to lay this at the bottom of the whole spiral, since the wool provides good drainage, while keeping the right amount of moisture*

→ *Then, you fill with what you have – thin branches and twigs, old straw, leaves, and so on. The top layer is made of topsoil, which should be approximately 10 cm thick. on the outermost layer, it may be appropriate to add a little fertilizer, since this area will house the most nutrient-demanding herbs*

→ *Remember to water and compress the bed material in time with it being put in. This is done so that the bed does not deflate too much while it settles.*

→ *The tiny little pond is then placed at the end of the spiral. Such a pond can, for example, be a concrete rhubarb leaf. If you have the opportunity, it is also good to place a larger stone to the south without casting shade on the herbs. Then it can store and add some extra heat when it is a little colder outside*

→ *Then to the most fun part – plant all the herbs!*

Intersection of a herb spiral

Sun traps

Simply put, a sun trap is a trap that collects a lot of sunlight. An effective sun trap has built-up earth slabs in the form of a horseshoe, with its' opening slightly twisted to the southwest. The size can vary, with everything from a small horseshoe bed in the garden to a large forest garden. In a garden, for example, you can make a small horseshoe bed to protect an especially heat-consuming plant. To strengthen the protective effect, one can then plant some bushes on the outside of the bed, and upon the bed itself some more durable plants that thrive and interact with the other plants.

In a vast sun trap, the embankments can be up to two meters tall, with the horseshoe having a width of 30–50 meters. Inside the horseshoe, plants that need extra sun and protection are grown. Outside of the embankments, plants that serve as a windbreak for the sun trap is planted. The sloping inner sides get an extra boost of sunlight.

At the back, the largest fruit trees are planted, which will also be able to provide some wind protection for the other plants. If the place is windy, one may have to start by making a windbreak outside of the fruit trees.

In front of the fruit trees, the slightly smaller trees are planted. Along the sides of the U, you can plan a mixture of small trees, trellised fruit trees or trellised berries – such as blackberries – or why not rosehip bushes. In front of, and on the inside of these, berry bushes of all kinds are planted. Perennial herbs are planted at the front.

The area now formed in the middle is excellent for slightly more sensitive plants.

A small sun trap for a single plant

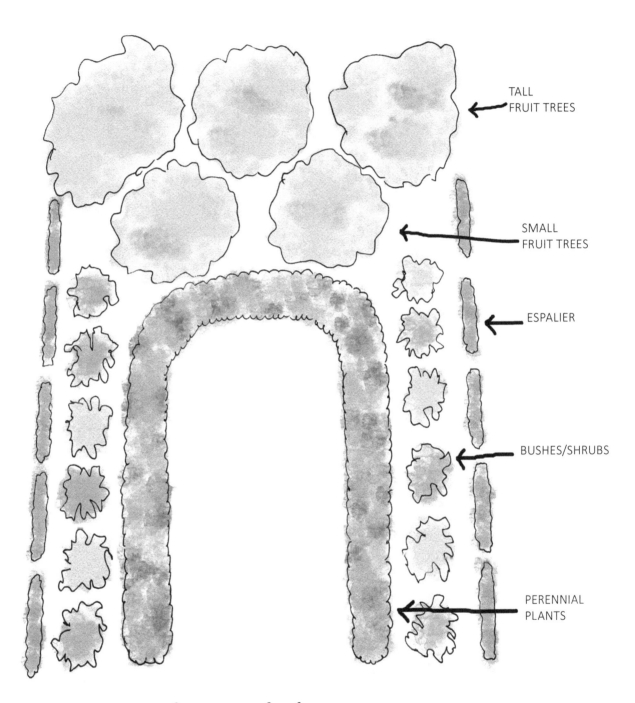

TALL
FRUIT TREES

SMALL
FRUIT TREES

ESPALIER

BUSHES/SHRUBS

PERENNIAL
PLANTS

A lagre sun trap for a large area

PEAS
BEANS
CLOVER

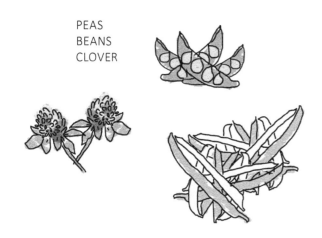

BRASSICAS
SQUASH
LEEKS

POTATOES

ROOT VEGETABLES
ONIONS
LETTUCE

The four year crop rotation plan

Crop sequence

There are several forms of crop sequences – or crop rotation, as it is also called. One way is that plants are planted sequentially on the same cultivation area within the season. For example, one can first grow lettuce, then dill; which also peaks quickly, and finally winter lettuce, which will finish growing late in the season. Another form of crop rotation is the one that runs over a number of years, and where the plants change their place of planting from year to year according to set rules.

In the past, this was a matter, of course. It was unthinkable to deplete the soil, and thereby expose the crops to diseases and insect attacks. One could still cultivate with nature. Since the 1950s, this knowledge has slowly but surely disappeared due to all the artificial fertilizers and other chemicals. Fortunately, it has started to swing back in to favour again, with old knowledge being dusted off.

It is not at all as difficult as one might think at first glance. By dividing the crops into four parts, which are then cultivated according to the 4-year crop rotation, you are already up and running. To keep track of what you grow and how it is progressing, it is excellent to keep notes.

Of course, you do not have to blindly follow these suggestions. As always, you have to adapt your cultivation to what you want to eat yourself. The basis of crop rotation is that plants with similar nutritional needs are cultivated together in each bed, while at the same time ensuring that the plants thrive on growing together.

Depending on how many different plants you want to grow, how large the area you have available is, and how good the soil is when you start your cultivation, you can choose between different amounts of years in your crop rotation cycle.

The benefits of crop rotation

→ *Reduced soil fatigue*

→ *Healthier soils*

→ *Reduced number of diseases*

→ *Fewer pest infestations*

→ *Optimal nutritional conditions*

→ *Different amounts of years in a crop rotation*

4-year crop rotation

Year / Bed 1 – Plants that provide nutrition

Year 1, you grow legumes and other green manure plants that feed the soil themselves, which means they do not need any extra nutrients. If the soil is impoverished, you can give a moderate dose of liquid manure or treatment of grass clippings in the spring. In late summer, when harvesting the peas and beans, the remains of the plants are worked into the soil.

If you grow pure green manure plants where you do not take a harvest, it is advisable to cute these down a couple of times during the season. Some of the cut off mass will also be an excellent cover material for the other crops.

→ *Beans*

→ *Peas*

→ *Green crop plants such as clover, vetches, and lacy phacelia*

Year / Bed 2 – Nutritionally demanding plants

Here you will feed the soil properly with fertilizer, covered with grass clippings, and water with nettle water, all done to ensure that the nutritionally demanding plants thrive. This group includes pumpkin plants, some onion plants, and cabbage.

→ *Brassicas*

→ *Turnips*

→ *Celery*

→ *Tomatoes*

→ *Cucumber*

→ *Squash*

→ *Melon*

→ *Maize*

→ *Leek*

→ *Garlic*

Year / Bed 3 – Moderate requirements for nutrition

Here you can add a little fertilizer in the form of mulching, but absolutely no new fertilizer – because these plants are not particularly demanding. Root vegetables grow for so long, and they utilize the mycorrhiza. Therefore, they manage with less nutrition. This includes root vegetables, lettuce, parsley, dill, and onions. Both root vegetables and onions will have a worsened durability and taste if there is too much nutrition in the soil.

→ *Carrots*

→ *Parsnip*

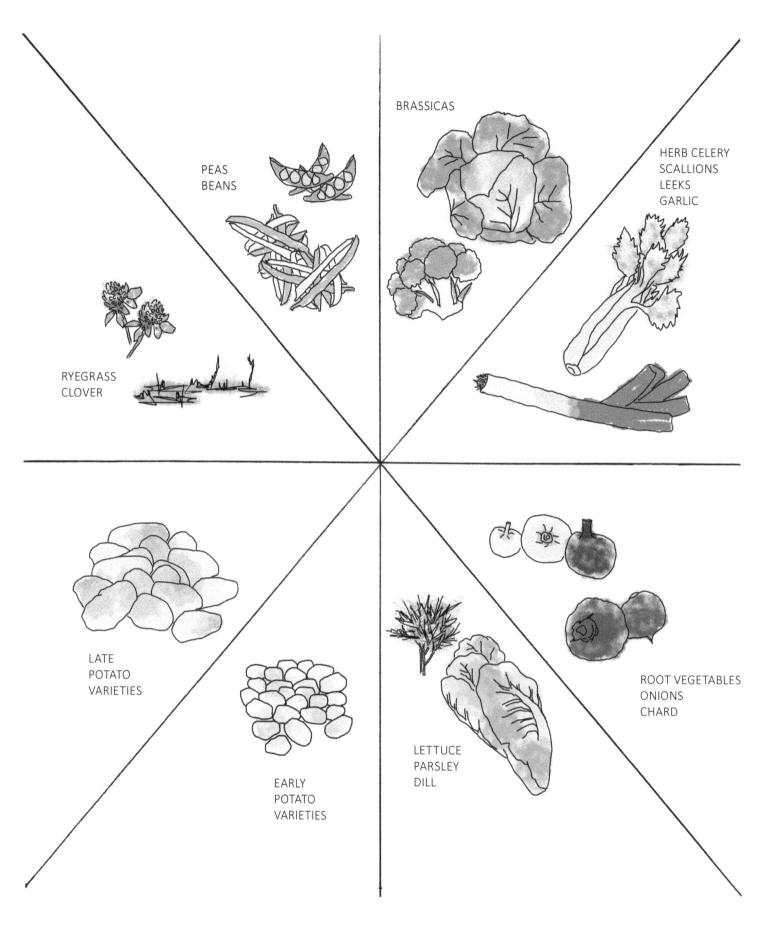

PEAS
BEANS

BRASSICAS

HERB CELERY
SCALLIONS
LEEKS
GARLIC

RYEGRASS
CLOVER

LATE
POTATO
VARIETIES

EARLY
POTATO
VARIETIES

LETTUCE
PARSLEY
DILL

ROOT VEGETABLES
ONIONS
CHARD

The eight year crop rotation plan

→ *Beets*

→ *Radishes*

→ *Spinach*

→ *Chard*

→ *Onions*

→ *Lettuce*

→ *Dill*

→ *Parsley*

Year / Bed 4 – Small nutritional requirements

The soil does not need to be fertilized, but feel free to a little cover material and plant, for example, broad beans between the rows.

→ *Potatoes*

5-year crop rotation

Year / Bed 5 – Green manure plants

As an alternative to the 4-year crop rotation, one year can be added per bed, while you only cultivate pure green manure plants the fifth year. The following year /bed (no. 1) changes only by removing the green manure plants.

8-year crop rotation

The easiest way to create an 8-year crop rotation is to divide the respective bed into two parts and expand the compositions as shown here. When you begin to find your footing, you can also rotate the plants within each bed. In this way, it takes many years before the same plant returns to the same growing area.

Much of the few

For example, if you want more potatoes than the space in the usual crop rotation allows, you can make a separate exchange system for the potatoes and a few other plants that you also want a lot of. For example, potatoes /green manure, beans/squash, pumpkin, leek/onion, lettuce, carrots and then back to potatoes.

Succession

Succession is another word for the slow change that is continually taking place within an ecosystem and which leads to the emergence of new forms of plant and animal communities. It is therefore vital that we understand and plan for how what we do in one place will change the site over time.

For example, newly planted fruit trees are quite small, and they grow slowly. For the trees to avoid competition during the establishment period, it is good to plant different one-year crops that cover the soil around the trees. This is good because weeds are kept away, soil erosion is avoided, evaporation is reduced – and, with the right choice of plants, nutrients are added to the soil.

When the trees grow larger, it is advisable to replace the annual plants with different perennial plants. When the trees are grown, some perennials can find it challenging to cope with the competition of the trees. Still, the cultivation area has nevertheless come to good use during the growing season of the tree.

Another example of succession is when an area has burnt. First, there will be bare soil, then annual plants, herbaceous perennials, shrubs and pioneer trees, and finally the burnt-down forest has become a real forest again. When nature manages this itself, each succession step improves the conditions of its successors.

Even in an annual planting, the succession can be used by, for example, sowing lettuce between slow-growing brassica plants. It works because the fast-growing lettuce is entirely harvested when the slower brassicas begin to require more space.

Or, you can then plant lettuce after lettuce, but of another variety. This works for most fast-growing plants.

Propagating plants

To propagate plants is in many areas forgotten knowledge since many people find it easier to go to the nursery and buy a plant there. However, with propagating, you create your own, brand new plant – not mention how much money you save at the same time. The only difference is that we have to equip ourselves with a little patience and let nature do most of the work for us.

It is relatively easy to take cuttings if only you are careful. Above all, it is vital to choose a nice and healthy mother plant; because the new plants will have the same characteristics as the mother. Then, there are several ways to propagate a plant in which different ways are better suited for certain plants than for others. The most common are:

→ *Winter cuttings*

→ *Summer cuttings*

→ *Offshoots*

→ *Root suckers*

But before I go through the different ways to propagate plants, there are a few other essential things to learn.

The right equipment when propagating

Cuttings are very sensitive to dehydration – mainly the summer cuttings – therefore you need to have everything you need ready in advance. It is also essential to ensure that all tools and the like are clean to reduce the risk of diseases.

For the rooting to go well, one needs a porous, fairly nutrient-poor soil, for example, seed compost, or two parts vermiculite and one part seed compost. If you then finish by laying a thick layer of sand on top of the soil, the sensitive root crown is protected.

You can plant in whatever you may have on hand, as long as it can be covered with some form of plastic or other relatively dense material. The cover should not lie directly on the plants, and an easy way to solve this is to support the plastic with steel wire, floral sticks or flexible pipes. Since high temperatures in the soil stimulate root formation, it is good to set the planter in a warm spot, for example, above a hot radiator. However, since the cuttings are sensitive, you need to be able to provide shade for them during hot sunny days. If you have proper support for the plastic cover, it is good to put some shade cloth over everything during the sunniest hours.

Your own root hormone

Another excellent way to give the root formation an extra boost is to use root hormone. But instead of buying an expensive one in the shops, you can easily produce one at home in your own kitchen. Do this:

→ *Pick this year's shoots from any salix (Sallows and willows)*

→ *Remove all the leaves*

→ *Cut the twig into pieces of 1–2 cm*

→ *Add 1 litre of water and leave for at least 24 hours*

→ *Put the cutting in the liquid and leave for 1–2 hours before planting*

Feel free to freeze the left-over liquid, it will keep for virtually forever.

Winter cuttings

Winter cuttings are last year's shoots from ligneous plants. Winter cuttings are best taken from late autumn when the plant is going into dormancy and until the end of January. The advantage of winter cuttings is that they basically manage themselves. This is how you do it:

→ *Prepare all equipment ahead of time, so that everything is ready to go when it is time to take care of the cuttings*

→ *Fill the pot with airy and moisture-rich soil consisting of, for example, two parts vermiculite and one part seed compost. Sand is an acceptable alternative to vermiculite*

→ *Select shoots that are nice and straight, and that are about 20 cm long. To know which end is up, it is good to cut the cutting with a straight cut. When you then shorten the top, you do it by cutting diagonally*

→ *Now is the time to ready the cutting. Remove 2 to 3 of the buds at the bottom. This part of the cutting should end up under the ground since it is here where the buds have been that the new roots will grow*

→ *Cut the top of the cutting so that 1 to 2 buds stick up over the soil. In total, you should have a piece with 3 to 5 buds on the finished cutting before it is planted*

→ *Now is the time to plant the cuttings. To stimulate the root formation, it is good to put cuttings closely together, and preferably along the outer edges of the vessel. In a pot that holds 1 to 2 litres of soil, 10 to 20 cuttings will fit. If you are planting a lot of cuttings, it can be easier to create a trench directly outdoors. Other than it being outside, the principle is the same as if you plant them in a planter*

→ *Water carefully. The soil should be moist, but should not be wet. Keep the soil moist, as the cuttings should never dry out. Similarly, if it is too wet, the cutting can rot*

→ *Place the pot either in an unheated greenhouse or in a sheltered place outdoors and cover with a thick layer of leaves to maintain the moisture levels*

→ *At the beginning of spring when the worst of the cold has released its grip, it is time to remove the leaves*

→ *Keep the cuttings protected from excessive sunlight, and keep the soil moist*

→ *Even if new buds are already sprouting in the spring or during summer, it is too early to repot the plants*

→ *First in August–September is it time to rehome them. Use planting soil mixed with some fine gravel to increase drainage capacity. Each cutting should have its own pot, 12–15 cm may be appropriate since it should not be too large. Then gently turn over the old pot with the cuttings and gently lift them out. Then, plant them at the same height in the soil as before*

→ *Continue to water and take care of as before. Protect with leaves for the coming winter. The next spring when the ground frost has thawed completely, the cutting can be planted in its final place*

Summer cuttings

→ *Some plants, such as different varieties of Salix (willow and sallows) are usually happy to root in water. Why; you can even plant them directly where you want them by making a deep hole that is narrow enough that it is only just possible to push in the twig. Conventional rebar works great for creating holes. Just as always, clean tools are essential. Only take cuttings from shoots that have sprouted that year*

→ *Since the cuttings dry out faster during the hot season, it is essential to take care of them immediately. Therefore, make sure that all equipment is prepared and ready to use. The best soil here is also an airy and moisture-rich soil, such as two parts vermiculite and one part seed compost. Sand is an acceptable alternative to vermiculite here too*

→ *The best time to do it is in the morning before it gets too hot. The ideal cutting is one that has a base that has started its' lignification, while the rest of it is green, soft and herbaceous. Take the cuttings from straight branches that are not in bloom, and that grows on the mother plant's sunny side. Keep in mind that the new plant becomes an exact copy of the mother plant. Therefore, choose cuttings with care. A good length of cutting is between 7–15 cm – depending on whether it is a large or small plant*

→ *Then, work in the shade or indoors. Start from the top of the cutting, and if it is very soft and thin, it is removed right above a leaf. Should any blossoms or buds have made it into the cutting, cut these off. The cutting should have at least two separate leaves or two pairs of leaves – but preferably 3 or 4*

→ *At the bottom, cut the cutting just below a leaf. If the shoot is long enough, it may be enough for several cuttings as long as there are enough leaves*

→ *Here, too, use a rooting hormone that the cuttings sit in for a few hours before they are planted. Another way to stimulate root growth is to cut down some twigs of Salix in the planting pot. These twigs quickly produce root hormone which then spreads in the soil and helps the cuttings take root more easily*

→ *Then, remove the leaves on the lower half/third of*

→ *the cutting and immediately place it in the soil. Here too, the cuttings can be planted tightly, but not so that the leaves touch*

→ *Cover the pot with plastic, or use a mini-greenhouse to maintain a steady humidity*

→ *The plants must not be exposed to direct sunlight, 70 to 80 per cent shading is ideal. The root formation works best at 18–24 degrees Celsius*

→ *Regularly check on the plants for aeration, watering, misting, picking out of mouldy leaves, etc.*

→ *Depending on the type of plant, it takes between 4 and 12 weeks before the cuttings root, but sometimes it can take longer*

→ *When it is clear that they have rooted, it is time to plant them one by one into pots and acclimate them.*

→ *Then they get left in their pots in a sheltered place over the winter. Cover the pots with a thick layer of leaves. It is also good to overwinter them in a greenhouse or in a ground cellar, as long as they do not risk drying out*

Propagate with offshoots

Another way to propagate shrubs is to use offshoots, which in principle means that you bend down a branch from an existing plant and help it take root. The best time for this is in the spring when the sap is rising, in early summer, and late autumn. However, it is not good to start an offshoot during the hottest and driest time of year.

Branch offshoots

→ *Choose a branch that is halfway through its' lignification – that is, a relatively young branch that sits near the ground*

→ *Put down soil that has been mixed with sand, or mix in sand into the existing soil at a suitable location below the selected branch*

→ *Scrape lightly at the bark on the underside of the branch with your nail, along the part that is to be in contact with the soil*

→ *Bend the branch down and press it into the soil you prepared*

→ *Put a stone or some other form of weight on it so that the branch is firmly attached to the ground. Alternatively, a U-shaped loop made of galvanized steel wire works well. Push the loop into the ground over the branch so that it is secured into place*

→ *After a few months – sometimes up to a year – roots will have formed, and the new branch can be cut off from the mother plant and replanted where you want it*

Snaking offshoots

Suitable for plants with long, flexible branches such as climbers or creepers. You do the same as above, but attach the branch to several places in the ground. Remember to scrape the underside of the branch at each attachment point.

Sprouting offshoots

Appropriate for plants with rigid branches that need to be renewed, for example, thicket shadbush, lilac, and dogwoods. Expect that the mother plant will not survive this treatment. This type of offshoot is made in winter.

→ *Cut the entire bush down to approximately 5 cm in height*

→ *As the new shoots begin to come up, they are covered with a sand/soil mix. Continue to fill with more soil until the shoots are about 20 cm tall. Remember never to cover the tips*

→ *In autumn, you can carefully shift the soil to see which shoots have rooted*

→ *These offshoots can now be taken apart and planted elsewhere*

End offshoots

Best suited for blackberries and raspberries. Since the most substantial concentration of root hormones is at the tip of a branch, this is the part you use.

→ *At the end of spring, the end is bent down into a shallow pit, which is about 10 cm deep*

→ *Attach the end offshoot with a U-shaped galvanized steel wire, preferably in several places so that the branch is firmly attached*

→ *Cover with a sand/soil mix*

→ *In autumn, carefully check to see which ends have formed roots. These can be cut off and replanted*

Air offshoots

Ideal for plants with hard, inflexible branches such as Magnolia or indoors on a fig tree. For this, sphagnum moss is needed. Up here in the north, we use "white moss" in our Advent decorations, which many think is the same thing – but remember that what we use for Advent is not sphagnum, but rather reindeer lichen.

Sphagnum moss you can buy either in a flower shop or in a pet shop that sells accessories for reptiles.

→ *Choose a young, healthy branch that has not completed its' lignification*

→ *Make a cut a few decimeters down the branch between two pairs of leaves. Note that the cut should only go halfway through the branch. Since such a cut naturally tries to close and heal quickly, the cut needs to be widened, either with a toothpick or some sphagnum moss*

→ *Brush on some rooting agent*

→ *Then, wrap the cut with pieces well-moistened sphagnum moss*

→ *Wrap and tie a plastic bag around the moss*

→ *Check regularly to make sure that the moss is always moist*

→ *Feel free to gently open the moss dressing to see if any roots have sprouted, but remember that it can take up to half a year*

→ *When the moss dressing is full of roots, it is time to cut loose the cutting and plant it*

Propagate with root suckers

Using root suckers is an effortless way to propagate a plant – provided that it is a plant that spreads through suckers, such as lilac. All you do is to dig up a root sucker carefully. It is crucial to make sure that you get a sufficient amount of roots before you cut off the sucker from the main plant. Then, plant as usual.

Propagate by splitting

Many of our perennial plants are OK to split after a few years. Either you pick up the whole plant and cut it in half and plant the different parts, or if you want to keep somewhere it stands, you can try to divide it there. Rhubarb and peonies, for example, are easily propagated this way.

Transport cuttings and shoots

If you get cuttings or other forms of propagation parts, it is essential to ensure that the plant and its roots do not dry out during transport. Split larger plants may need to stand in a bucket with some water. Others smaller parts can be wrapped in newspaper that is very wet, and transported in a plastic box. Rather too wet than too dry – for a dehydrated plant never recovers.

A food forest that mixes the natural and the wild

With the forest as a role model

Throughout history, man has interacted with nature, and of course, with the forest. Today, we have begun rediscovering several of the old methods of natural interplay. Ways that utilize cultivation areas more efficiently while becoming more ecologically sustainable.

Intercropping with trees

The methods are based on the fact that trees are in several ways intercropped with other crops and on pastures. Here, trees are used for several different purposes, such as giving crops, nourishing other plants and soil, for animal feed, wood, timber, as protection against pests, and as wind and sun protection. Thanks to the trees' many functions, these forms of intercropping create greater biodiversity and a more sustainable ecosystem at that place. In this chapter, I will describe some different methods that are in use around the world.

Silvoarable

The method "Silvoarable" means that trees are planted as windbreaks between different cultivation zones. Many varieties of trees are used - trees for timber and wood, for harvesting fruit and nuts, for pollarding fodder, or a nitrogen-fixing purpose. The tree varieties are chosen depending on what you want to achieve. The trees can be planted either in rows or in clusters.

Silvopasture

Silvopasture is a method that builds on trees being planted in rows or clusters, around or on forage lands and pastures. Tree varieties and features are the same as with Silvoarable.

Forest farming

Forest farming is a method that means integrating the crops into the already existing forest landscape. Common crops grown here include mushrooms, nuts, root vegetables, fruits, berries, plants for bees, herbs, medicinal plants, and edible flowers - as well as trees that can produce edible crops. This way, you create both a food forest

Silvopasture

and a forest for timber, wood, and woodchips. Integrating different animals into the system is ideal as well.

Forest garden

Forest gardening is nature's version of a permacultural system, which manages itself in almost every case.

The difference between the forest as mentioned earlier cultivations and these forest gardens is that in a forest garden, both forest and crops are planted from scratch.

In the forest garden, all conceivable forms of fruits, berries, vegetables, and herbs are grown side by side in the same type of ecosystem as in the natural forest. We can create our food forest by learning how the natural forest is constructed.

A natural forest consists of many different vegetation layers, which are:

1. *Overstory (canopy) tree layer - large and medium-sized fruit trees*
2. *Understory tree layer - dwarf species of fruit and nut trees*
3. *Shrub layer - fruit and berry bushes*
4. *Herbaceous layer - herbs and perennial vegetables*
5. *Vertical layer/Vine layer - climbing and trellised plants*
6. *Ground cover layer - creeping groundcover plants*
7. *Root level - root vegetables*
8. *Mycorrhiza level*
9. *Water level*

Plants within a forest garden

When choosing plants, it is essential to select these with regard to both the overall climate, the local conditions, and the needs of the plants and their interactions. This will increase nature's own ability to help create the right environment. An adequately designed forest garden, with its immense biodiversity, has a broad and wide range of different habitats that in turn create many opportunities for abundant wildlife - both below and above ground.

Creating a forest garden

Creating a forest garden takes time and requires a lot of knowledge, but I will guide you through the basics.

Planning so that the forest garden contains many different species - all with different conditional needs and that become harvestable at different times - not only provides food for a longer part of the year, but also strengthens its' protection against bad weather, diseases, and pest attacks.

In a forest garden, many jobs take care of themselves.

Here, the nutrient supply occurs by all kinds of green plant parts falling to the ground, which then decomposes and is dealt with by the soil life. The absence of bare soil prevents soil erosion and evaporation. This environment also favours all forms of animals - from the tiniest insect to the largest moose - all of which add to the overall diversity. Additionally, all the nutrients in the soil come to use, since different plants get their nutrients from different depths.

Building a forest garden requires a lot of planning before you can start physically building. It is essential to have the right growth in the right place, both in the short term and in the long term. It is good to plan so that the same species do not end up next to each other to reduce the risk of spreading diseases.

The easiest way is to start by placing out the big trees by taking into account which thrive together and which ones need each other for pollination. Then, it is necessary to intersperse small trees and bushes in the same way, while taking into account which can tolerate shade, and which needs sun.

This may mean that you need to move some of the larger trees, or that some areas may be better without these. The next step is to plan for herbs and groundcovers. There are undoubtedly some trees that will work as support for climbing plants.

An excellent way to get a lot of space on a small surface is to plant in waveforms, as is often the case in Permaculture. It not only allows for more plants on the same surface, but it also provides more sun for ach plant.

A drawing is a must in order to succeed in planning - as there is a myriad of details to take into consideration. A good tip is to use small, loose markers for different things that you can then move around on the drawing.

Creating a forest garden requires quite a lot of work during the first few years to ensure that everything gets a good start. After that, there will be proportionately little work considering the abundant harvest. Over the years, this and that may need to be replaced or supplemented. Still, overall, a forest garden is easy to maintain.

A U-shaped forest garden

The U-shape that I went over earlier is a popular and often used design in forest gardening as well. The intended surface is designed as a large U, where the opening is placed slightly towards the southwest so that you can utilize the sun's energy to the fullest.

Different layers of a food forest

1.   OVERSTORY (CANOPY) TREE LAYER - LARGE AND MEDIUM-SIZED FRUIT TREES
2.   UNDERSTORY TREE LAYER - DWARF SPECIES OF FRUIT AND NUT TREES
3.   SHRUB LAYER - FRUIT AND BERRY BUSHES
4.   HERBACEOUS LAYER - HERBS AND PERENNIAL VEGETABLES
5.   VERTICAL LAYER/VINE LAYER - CLIMBING AND TRELLISED PLANTS
6.   GROUND COVER LAYER - CREEPING GROUNDCOVER PLANTS
7.   ROOT LEVEL - ROOT VEGETABLES
8.   MYCORRHIZA LEVEL
9.   WATER LEVL

Extending the growing season

Here in the north where the growing season is short, it is precious for us to be able to extend the growing season. The contrivances we take are many - with everything from starting seeds indoors so that the plant gets a sufficiently long growing period to carry a harvest, to building greenhouses where we can grow things no matter the season.

Starting plants indoors

Starting plants in your house gives hope for a new spring and summer. In order to grow and harvest some plants here in the north, it is necessary to start them indoors first. Some other plants enjoy it as well since it gives them a better chance to survive once they get put outside. However, it is quite space-consuming to do this on a larger scale, and you need to have a genuinely bright place for them. If you do not have enough light, it is possible to solve this with different lighting setups. If you are en-

tirely unable to start them at your place, you may be able to house your plants with someone with better conditions.

Cold frames and hotbeds

A cold frame is a small, low greenhouse. With a simple cold frame consisting of a solid frame and a transparent lid, you can extend the growing season by about a month during both spring and autumn.

The longer side of the cold frame should be turned

Cold frame

slightly to the southwest, and the rear wall should be higher than the front to get the most out of the sun. The tilted lid also makes snow and rain slide off easier. Materials for the frame can be concrete or wood, and the lid any material that lets the sunlight through accurately. It is common to use old windows. To make the climate in the cold frame even better, you can insulate it with soil, leaves, or straw around the outer walls.

If you want to extend the season further, you can instead make a hotbed. To increase the heat in the bed, one begins by laying out a thick layer of fresh horse manure mixed with lots of straw, The thicker you make it, the warmer it will be. The next step is to pack in the manure firmly - easily done by walking on it - and water it with hot water, preferably with added urine to speed up the heating process. You can then set the frame in place, and put the lid on. You then barricade the frame with straw all around it. Soon the heat will start to rise and then begin to fall again. When the temperature has come down to about 30 degrees Celsius, it is time to add the topsoil and start growing things!

Keep in mind that it may get sweltering underneath the glass when the sun's rays start to warm. Therefore, you need to keep a close eye on the temperature. Photosynthesis gradually decreases when it gets to be above 25 degrees Celsius, which also reduces growth and production. The further along in spring you come, the more often the cover needs to be lifted during the warmest hours. Another way is to cast shade on the beds with, for example, reed mats or old rugs.

With the help of both cold frames and hotbeds, you can get fresh leafy vegetables much earlier in the year. A hotbed can provide its' first harvest as soon as in February.

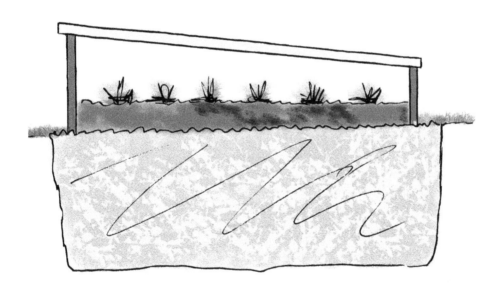

Hotbed

Greenhouses

For many, the dream is to have a greenhouse. It can be anything from a small greenhouse built from old windows next to a southern wall, to a large garden greenhouse. The range of greenhouse kits is vast, but they cost quite a bit. But if you are good with tools, you can build one yourself by putting up a wood frame that you then cover with construction plastic. Another option that is cheaper than regular greenhouses is the so-called arch greenhouses, which have steel frames covered with plastic tents.

All greenhouse frames must be anchored well into the ground so as not to fly away as soon as the wind picks up. Greenhouses made from glass are in themselves quite heavy. For these, you only need to screw the body properly into the underlying base. When it comes to plastic greenhouses, you can bury the edges in the ground. Still, a better way is to build a rail inside the greenhouse against which you then attach the plastic. If the place is windy, you may need to create a natural windbreak out of plants to protect it.

I will not go through the ups and downs of growing in greenhouses since there is already a lot of literature on the subject. However, there are several good ways to combine a greenhouse with another house, which less often brought up. Thus, here it is.

Greenhouses combined with outbuildings

Connecting a greenhouse with another building has many advantages, provided that you put the greenhouse

A chicken coop and greenhouse combined

on the side of the house where you get the most sun. The most significant advantage is that you save a lot of energy - both in the greenhouse and in the adjacent building. Having a glass wall to the north gives no merit as far as the sunlight goes, rather this only leads to energy losses.

Depending on what you have in the other half of the building, you can even create an additional energy supplement for the greenhouse. The previous page shows a way to combine a hen house and greenhouse. You can also combine the greenhouse with other spaces, such as workspaces for planting, sowing, propagating, and for storing your harvest. You can also have water tanks where you collect the rainwater from both roofs. The options are many, so you have to think about what you need.

Greenhouses against a house wall

An excellent way to utilize the joint effects of a greenhouse next to another building is to build the greenhouse along the sunniest side of the home. This will have a climate equalizing effect inside the house - not only during hot summer days but also during cold winter nights.

The energy consumption will be smaller for both the home and the greenhouse. At the same time, you get a cultivation area directly connected to the home. If it is possible, you can let the greenhouse cover the entrance to create an arctic entry, which saves even more energy during the cold season.

Greenhouse for hot water

Another interesting solution is where the greenhouse can help to heat water for various purposes. Since no plants like being watered with cold water, you can avoid this by storing the water in a barrel or tank inside of the greenhouse.

Another system is to heat your water - at least during summer - using a small greenhouse as in the picture below.

A greenhouse set against the wall of a house

Getting a greenhouse

Once you have decided to invest in a greenhouse, there are many more factors to consider than just the cost.

→ *How big does it need to be?*

→ *Do you buy a kit or build from scratch?*

→ *What weather factors need to be taken into account? For example, how much weight does the roof need to be able to withstand in winter?*

→ *Should it be glass or plastic?*

→ *Which building materials are best suited? Keep in mind that stone stores heat better than wood.*

→ *Which plants should be grown there?*

→ *Should it be used for something more than cultivation?*

→ *Will you grow in beds or in planters and pots?*

→ *Is extra lighting needed?*

→ *How do you handle the ventilation?*

→ *How will the irrigation be set up?*

→ *How do you manage the nutrition the plants need?*

A greenhouse for hot water

Self-sustainability - How much you need

Self-sustainability is becoming increasingly popular, and many are thinking about what you need and in what amount to become fully self-sufficient. This is not an easy question to answer since many factors come into play. This chapter, therefore, provides a more general picture of what is needed.

Finding information on how much you need to grow to become self-sufficient when it comes to vegetables made me a right detective since most of the information available is about how to grow - not how much you need to grow.

To at least give a clue as to how much you might use, I have compiled information where the numbers mainly come from a thesis from 2010. The calculations are based on a mixed diet of both vegetables and meat, where all food is eaten at home. Specific crop loss and that the crops are located up here in the north (specifically, in the middle region of Sweden) have also been taken into account.

In total, this would mean that you need the following for a family of two adults and two children:

→ *334 kg potatoes (including rice and pasta)*

→ *130 kg of root vegetables*

→ *195 kg more nutritious vegetables*

Annual consumption of vegetables

Annual comsumption of vegetables		Average consumption		Varav	
Produce group	Produce	grams/week	kilos/year	grams/week	kilos/year
Potatoes, rice and pasta		1 607,5	83,6		
Root vegetables		622,5	32,4		
of which:	Rutabaga/Swede			90,0	4,7
	Carrot			375,0	19,5
	Parsnip			67,5	3,5
	Celeriac			45,0	2,3
	Beetroot			45,0	2,3
Vegetables		937,5	48,8		
of which:	Onion			180,8	9,4
	Leek			45,0	2,3
	White cabbage			198,0	10,3
	Broccoli			90,0	4,7
	Cauliflower			40,0	2,1
	Peas			85,0	4,4
	Beans			40,0	2,1
	Spinach			85,0	4,4
	Frozen mixed vegetables			40,0	2,1
	Peppers			113,0	5,9
	Maize			23,0	1,2
TOTALT		**3 167,5**	**164,7**		

The land needed for this is, according to the thesis, just over 500 sqm of cultivated land.

How much land you actually need, and how much you need to grow is determined by the family's dietary choices, how much they eat at home, and what farming conditions exist on the site.

In the thesis, there was also an attempt to calculate the time required - that it should suffice with 25 minutes a day elapsed over six months to grow all this. However, I don't believe this is reasonable since there were a few too many assumptions combined with a lack of figures on how long manual cultivation takes. My own experience tells me that it is probably going to be 2-3 hours a day throughout the growing season. Of course, the time required depends on what is grown and with what methods. The time requirement will always look different; some days with no work at all, and some days with many hours of work.

In addition to all the greens, you need a lot of fruits and berries. Of course, how many shrubs and trees you need also depends on how often you eat at home and what opportunities you have to take care of and store fruits and berries over the winter.

The estimate below is, therefore, also a clue as to what might be needed. The number of plants is based on that the family can store the harvest, and that most meals are eaten at home.

Which fruits and berries you can grow will depend on in which zone you live.

→ *4 apple trees*

→ *2 pear trees*

→ *2 plum trees*

→ *3 cherry trees*

→ *20 raspberry plants*

→ *6 currant bushes*

→ *6 gooseberry bushes*

→ *70 strawberry plants*

In addition, you may want to both add or replace some of these plants with, for example, blackberries, blueberries, sea buckthorn and peaches.

Household animals

A chapter on keeping pets may not be what everyone expects from a book about how to garden. But for your land to become similar to nature's ecosystem, animals must be allowed to do their part.

To give some inspiration along the way, I have made a brief compilation of the most common animals when it comes to self-sufficiency.

The animals are vital since they are useful in many different ways; such as processing the land they walk on, providing fertilizer to our cultivations, and providing food of various sorts. At the same time, animals give us love and warmth when we spend time with them in their daily care.

Animals require both responsibility, respect, and effort, but they provide plenty in return. So before you acquire your animals, you need to know what the different kinds need and how well they suit what you can offer.

If you are a beginner, it is good to start with small, relatively easy to care for animals. That's why many people start with chickens. Another relatively easy animal is ducks, but they must have their duck pond. The next step is often a few sheep - larger and requires a little more of everything. Then maybe it's time for a household pig and then finally the big step - cows. They are probably the most demanding of all our domestic animals, especially if the family wants milk.

Here in Sweden, we have the Swedish Board of Agriculture that requires that the animals are looked after twice a day. Remember to check your country's laws.

Chickens

Chickens are, as previously stated, often the first animal many people get. They are durable, productive, relatively easy to handle, and require proportionally little space. Chickens provide a reasonably good return compared to the work put in - by providing both eggs, meat, and fertilizer. Chickens are also good at processing the soil, and they require relatively little care.

The more you socialize with them, the more social they become. If the chickens learn that you have something tasty with you, you will be met by a bunch of running, happy chickens as you approach the chicken coop.

If you live in a more densely populated area, it may be advisable to refrain from having a rooster. In some municipalities (here in Sweden at least) it is even prohibited to keep roosters in more densely populated areas. However, you are allowed hens in most municipalities.

Geese

Geese are primarily kept for their meat since they do not lay that many eggs. Geese are good grazing animals as they are herbivores. Also, they can utilize pastures that are too small for, for example, sheep. They also do not require much work and are easy to fence in, as they do not fly or jump very well.

Geese live in lifelong relationships and can die of heartbreak if one of them dies. Some geese are quite noisy, and they make a mess due to their stool being larger than smaller birds.

Geese can grow fairly old - up to 10-15 years. With age, they can become quite stubborn and are happy to act as "watchdogs" around your home. Geese are also more intelligent than other poultry, so anyone interested can teach their geese a few tricks.

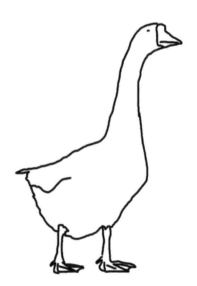

Ducks

Ducks are kept for their meat, and they too are grazing animals. Most ducks are sweet and somewhat talkative birds. They are typical herd animals, and the behaviour follows suit. Completely unprovoked, they can set off at full speed uphill and down dale to knock over everything in their path. They can be a bit messy, so you may have to muck out their winter home pretty much every day. Ducks have a more mixed diet than geese, and they will not only gobble up the grass but also insects, worms, snails, and slugs.

Muscovy ducks

Muscovy ducks are also kept primarily for their meat. It is said to be the best meat among domestic poultry, at its best if they are allowed to go outside and eat their natural diet.

Muscovy ducks are birds with personalities and are happy to be around their owners. However, they can be difficult to fence in since they both climb and fly. They are large birds and therefore leave quite a lot of droppings.

In areas with a lot of Spanish slugs (not-so-affectionately called "killer slugs" in Swedish), they have proven to be a great asset by happily eating large amounts of these.

Although they are good egg-layers and good at taking care of their young, they are not as popular as you might wish. Maybe this is because they are considered rather ugly, but they are well worth having.

Rabbits

Rabbits are an underrated animal from a food perspective as they are usually bought for the sake of the children. The best rabbit variety here in the north is the "Gotlandskanin" - a rabbit breed from Gotland - since it is low maintenance and very social. A good international breed is the Champagne d'Argent rabbit, also very easy-going, docile, with a fair amount of meat. Keeping rabbits is done for the meat, pelts, and potential income from sold kits. The meat is not very tasty in itself, but the more varied diet of different grasses and herbs, the better. You can also try to give them sticks of juniper, which is supposed to provide the meat with a more "wild" taste.

The easiest way to keep a few rabbits is to use the classic netted rabbit cages with a house at one end. The cage should also have netting on the bottom as rabbits like to burrow. However, the rabbit cages require some work as they must be moved frequently so that the animals have fresh grass to eat at all times.

If you want a lot of rabbits and have space, they are happiest if they are allowed to be outside in a larger fenced area with a house for protection. Of course, they may decide to dig a tunnel and escape, but they usually stay close to home. If you equip yourself with a net, it is easy to retrieve the escapee. It is best to shut them in at night to protect them against foxes. The rabbits can survive outside year-round as long as they have proper wind protection with dry berths. Rabbits do not like living in heated rooms or spaces.

Bees

Beekeeping has become increasingly popular in recent years as more and more people realize the advantage of having your own pollinators while also getting honey out of it.

There is a lot to think about when beekeeping, such as regular checks of the bee communities to ensure that the bees are healthy and to counteract swarming. Depending on how many bees you have, collecting honey can also be quite time-consuming. They require specialised care, so taking a is beginner's course is ideal.

Goats

Goats are kept for their milk, meat, and sometimes for their pelts. The goat has a strong scent, and its' milk and meat has a specific taste that not everyone likes.

Goats are very intelligent, inventive and easily taught. If you spend a lot of time with them, they will become incredibly tame, why - they can even be housebroken. Goats love to climb and jump, and in combination with their curiosity, it is common for them to try to escape. Therefore, electric fencing is required to keep them in. Even so, there is no guarantee that they will stay put.

Goats can be outside all year round when they have access to a wind shelter with a dry berth. However, if the winter is harsh, you have to have a stable for the goats since they have a thinner undercoat than sheep and therefore can not handle the severe cold.

If the family has children, goats can be a better choice than sheep since the goat does not guard its offsprings in the same way as sheep does. Goats are also the world's best brush cutters. They mainly eat shrubs and need to eat both branches and bark in addition to grass and herbs. Goats also have a natural but less appreciated affinity for tasting just about anything they see to determine if they can eat it, including clothes and paper.

An alternative to common goats is pygmy goats, which are even better with children. However, these do not provide any meat, just milk.

Sheep

Sheep are kept for their meat, milk, wool, and pelts as well as for their excellent grazing ability. They are easy-going, social, and kind.

If you have a smaller herd and spend time with your sheep daily, they can become very tame. If you are planning to have lambs, it is best to start with breeds that have easy births and are good at taking care of their young, such as the Swedish Gute Sheep or and the British Milk sheep.

Sheep do not require much but need just as much space in the form of pastures. Most sheep breeds can be outside all year round when they have access to a wind shelter with a dry berth. In short, sheep are an excellent domestic animal that is relatively easy to care for, but taking a course is still recommended.

Pigs

The classic household pig has gained a new following when it comes to self-sufficiency. The pigs provide plenty of good meat while being excellent tillers.

Pigs are sweet, very intelligent, and easy to tame. They will happily learn the odd trick and will follow you on the evening walk if you let them. But remember that pigs are very quick-footed despite their clumsy appearance. Therefore, young children should not be left unattended with them.

Keeping a couple of pigs until the autumn slaughter is quite easy. What is needed is a substantial fence, possibly with electricity, good access to water, and perhaps extra food - preferably in the form of household waste (compost) and a draft-free, dry berth. Keeping pigs all year long requires more effort, but they are still a relatively easy animal to keep. Some breeds can be somewhat aggressive when they mature, however.

Cows

Last but not least, cows. It is not an animal for the beginner, at least not without first acquiring the proper knowledge. Many say that cows are the household animals that give the most back, and not just milk and meat. Cows have great personalities, and they are both wise and social. Therefore, they often become dear family members.

Cows are large animals and cost accordingly. They require large spaces, both indoors outdoors. You also need space to grow winter feed and to deal with the manure.

It is quite a lot of work, and if you choose to have dairy cows, they need milking once or twice every day.

The easiest way to start is with a couple of older cows that are already used to being handled in, for example, milking. If you want to rear calves, then these naturally learn from the adult animals how the daily routine works.

About the author

Raised on a farm on the island of Gotland, Sweden, where her family grew potatoes, strawberries, and other vegetables by hand - led to Eva being very tired of farming when she moved away from home at 18.

The farmer in Eva never gave up

Eva then proclaimed that she would never live in the countryside again! But at 25, her attitude had already softened. The lack of fresh produce right outside the door made itself known. So in the muddy lawn of her terraced house, Eva planted the first potato. And at 30, Eva left the city and has since lived in the countryside in different places, with varying amounts of gardening space. Just past 40, Eva got her first greenhouse. And not a small one either - a whooping 40 square meters. It was joy, pure joy; now she could grow more produce, with an extended growing season.

Her interest in food and the quality of it awakened as Eva and her family started a restaurant in 1998. The guests would be offered the best ingredients available, some home-grown, others from local farmers. What could not be obtained locally would at least be Swedish in origin. Eva remembers that they wrote in their menu that "We serve food from Gotland, and in the worst-case - Swedish".

Some tourists probably thought it was a bit cooky, but not the Gotlanders. Another memory from that time is the numerous discussions had with the meat wholesaler, who for a long time insisted on delivering Danish pork when the Swedish was out of stock. Each time Eva pushed back; she only wanted Swedish meat, so the Danish meat had to go back. Nobody understood what she was on about. It is only now in recent years that more and more people are beginning to understand what Eva meant back then.

A few years ago, Eva came into contact with something called "permaculture" and became curious. The more she read about, the more interested she became. So much so that Eva studied to become a Certified Permaculture Designer. Once she started the course, her view of nature and cultivation changed forever. To think how much easier everything is when people take a step back and learn from nature - and not only when it comes to gardening!

In the beginning, not many people knew what Eva was talking about when she began to talk about permaculture in a garden setting. Still, over time, more and more curious questions started to pop up.

Several of her friends asked if she knew of any courses for someone who wanted to learn how to use a permacultural approach in their gardens. She did not, and it turned out that there were none anywhere close-by.

So Eva thought, how hard can it be... and promised her friends to arrange a course.

Then reality hit - when you want to teach, you need course material! Once she started writing, Eva soon realized how much knowledge she wanted to convey. Knowledge which, if you only know about it, makes growing so much easier. The result of all that writing eventually became the book you hold in your hands.

Thanks from the author

I want to thank my family and friends who have supported and believed in me throughout the process of writing this book. In particular, a big thank you to Madelene Dahlberg for doing the layout, cover, and the translation from Swedish to English.

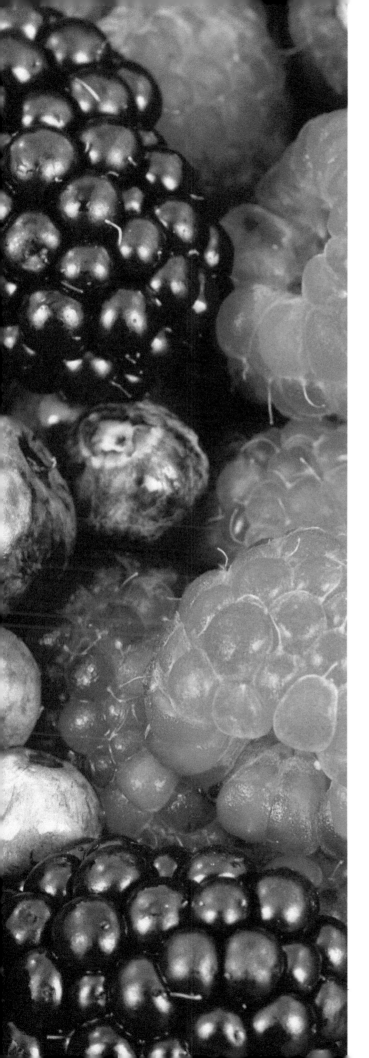

Appendix

In this last part, I have collected lots of information that I hope you will enjoy.

All information should be regarded as general advice. Before using them, I recommend that you read about the different plants you are interested in to find out if they fit your conditions and desires, and to find out if any plant needs special treatment to be eaten safely, or if any part of the plant is inedible.

The sources span from the Swedish Board of Agriculture, via facts from nurseries to both Swedish and foreign reputable growers.

Parts of the appendix

Green manure plants

Plant family	Genus	Example	Nutritional needs/Soil	Note
LEGUME FAMILY Fabaceae		Pea, bean, clover, lupine, lucerne, broadbean, sweet clover, fodder vetch, hairy vetch	These plants need no extra nutrition unless the soil is impoverished.	
BORAGE FAMILY Boraginaceae		Comfrey		
	Phacelia	Lacy phacelia	Does not provide as much nutrition, but improves the soil by increasing the humus content and the micro life becomes more prosperous.	
GRASS Poaceae	Festuca	Red fescue		
	Lolium	Perennial and Italian ryegrass.	Does not provide as much nutrition, but improves the soil by increasing the humus content and the micro life becomes more prosperous.	French ryegrass is of a different genus.
ASTERACEAE	Calendula	Pot marigold (calendula)		
	Helianthus	Sunflower		

Plants for companion planting schemes

Plant	Thrives with	Thrives Not with	Soil, Environment and Sowing	Harvesting	Problems and Solutions
Ajuga reptans	Rose				
Anise	Coriander, cabbage. Anise and coriander help each other grow better.				
Aniseed	Good for bees				
Appletree	Barberry, elderberry, rowan and guelder-rose can reduce bird infestation on apples.	Walnut			
Asian leafy vegetables	Fragrant herbs, onions, celery, potatoes. Applies to the cabbage varieties - (other varieties reasonably easy to attack).	Common beans, strawberries	The cabbage varieties work best with late sowing to avoid premature flowering. Other family varieties are not as sensitive. Want nutritious soil and even moisture. Sow 1 cm deep, sparse in batches. The best harvest is usually sowed in July.	Start picking the leaves early They grow quickly and easily bloom in summer...	The cabbage varieties: Sprinkle wooden ash or stone flour on and around the plants repeatedly against flea beetles and cabbage caterpillar. Co-culture to reduce attacks.
Asparagus	Basil, dill, coriander, parsley, parsley root, tomato.	Scallion			
Aster	Attract many beneficial insects and predators.				
Autumn Aster	Attract beneficial insects during late summer and autumn. Feel free to complement with sunflowers and coneflower.				
Barley	Also, see cereals	Poppy. Barley, the year before potatoes, can give potato scab.			

Plant	Thrives with	Thrives Not with	Soil, Environment and Sowing	Harvesting	Problems and Solutions
Beet	Runner bean, turnip cabbage, rutabaga, red onion	Tomato, peas	Kohlrabi, peas, and beans can be sown the year before. Can be grown on most soils. Avoid rigid clay soils. At least 4-year crop rotation. Best in humus-rich and sandy soil. The nutritional need is moderate and too strong nitrogen fertilization gives hairy, overgrown, and bad roots. To avoid potassium deficiency, give some wood ash or stone flour.		
Beetroot	Bush bean, dill, cucumber, cabbage (kohlrabi), onion, radish, lettuce, garlic.	Corn, potato, leek, spinach, climbing bean, tomato.	Can be grown on most soils. Avoid stiff clay soils. At least four years of crop rotation. Best in humus-rich, sandy soil. The nutritional need is moderate and too strong nitrogen fertilization gives hairy, overgrown, and bad roots. To avoid potassium deficiency, give some wood ash or stone flour. Kohlrabi, peas, and beans can be sown the year before. The seed is a bit demanding, sow preferably in several rounds for a better harvest.	For storage, harvest before the intense frost nights begin. Leave a few cm of the tops and do not polish off the roots.	
Birch		Potato			
Black radish	Bush bean, strawberry, chress, cabbage, onion, chard, carrot, parsley, tomato, pea.	Cucumber, hyssop, spinach, squash, tomato.			
Bramble	Common tansy. Favours both growth and taste				

Plant	Thrives with	Thrives Not with	Soil, Environment and Sowing	Harvesting	Problems and Solutions
Broadbean	Leaf celery, cauliflower, dill, cucumber, cabbage, carrot, potato, spinach, white cabbage, peas. Potato protects against aphids.	Broadbean, onion, garlic. Broadbean does not thrive in monocultures.			
Broccoli	Dill, onion, mint, geranium, leek, rosemary, radish. Dill and mint provide better growth. Rosemary against cabbage flies. Geranium against cabbage worm. Radishes against lice.				
Brussels sprouts	Onion. Increases growth and harvest.				
Bush bean	Cauliflower, dill, cucumber, strawberry, coriander, spice herbs, cabbage, chard, early carrot, potato, rhubarb, rosemary, celeriac, radish, black radish, beetroot, lettuce, celery, summer savoury, spinach, scorzonera, tomato, corn. Promotes the growth of cabbage plants. Corn and squash are the best co-grow plants.	Fennel, onion, leek, scallion, white cabbage, gladiolus.	Soil improvers, green manure.		
Capita group	Basil, broadbean, tarragon, hyssop, anise, dill, thyme, celery, nasturtium, oregano, wormwood, sage, rosemary, tansy.	Strawberry, climbing beans, white mustard, marigold, common rue.			

Plant	Thrives with	Thrives Not with	Soil, Environment and Sowing	Harvesting	Problems and Solutions
Carrot	Cauliflower, broadbean, bush bean, chive, onion, leek, lettuce, tomato, garlic, herb spices, radish, black radish, turnip, chard, scorzonera, parsley rot, sage, tomato, garlic, pea. Leek is the best co-grower for the carrot. Stimulates pea growth.	Dill, parsnip, potato, rosemary. The carrot fly can attack the parsnip.	Prefer humus-rich, sandy soil with a balanced pH. Works well on other light soils. Consume a lot of potassium. Too much nitrogen produces low durability, cracked roots, unnecessary leaves. Vital is a light and airy plant site with moist soil. Nettle water strengthens the plant. During summer and autumn, you should cover the plants but be careful as the carrot is susceptible to damage. Sow 1 cm deep and not too dense. The soil should hold at least 7 C. The seeds can be germinated by placing it in lukewarm water for a day and then allowing them to dry again. Water the soil before sowing. Avoid it becoming a hard soil surface as the sprout then becomes challenging to get through. Not grown more often than every 4-5 years in the same place. Risk of fungal infection and infestation	The leaves should still be green and fresh - not discoloured or withered. Withstand several degrees of frost but allow them to thaw completely before shooting. Handle carrots with care when harvesting, as bumps degrade sustainability. Leave some cm of the tops. Best stored in sand, dry leaves, peat, or sawdust in basements. If you have perfect storage space with 0-1 plus degree and high humidity, they stay only in bags.	The carrot leaf flea and the lice that cause curling of the carrot leaves as well as small inedible roots. The yellow-white caterpillars of the carrot fly making tunnels in the roots. Avoid moving the leaves as much as possible to avoid spreading the scent. Advantageously clean in wet weather. Powder with stone flour, even fresh sawdust, conifer, fern or tansy leaf, snuff or soot in the row can help.
Cauliflower	Beans, carrot, rhubarb, marigold, black radish, lettuce, celery, spinach, marigold, tomato, peas. Beans, onions, and celery give a better taste.	Cabbage onion, potato, garlic. Not after spinach			

Plant	Thrives with	Thrives Not with	Soil, Environment and Sowing	Harvesting	Problems and Solutions
Celeriac	Bush bean, bean, cabbage, kohlrabi, onion, lettuce, tomato. Be careful about the crop rotation because of several plant protection problems. Mixed culture is good for celery, which thrives well with beans and onions, but also with tomato and cabbage. Particularly successful is the combination of celery and cabbage. Celery helps to keep; for example, cabbage flies away because of its strong scent.		Grow early indoors. Light sprouting, therefore, it is sown shallow 2 mm already in Feb-Mar 10-12 v before transplanting.	Grow best in autumn and can withstand several minus degrees. Defrost appropriately before harvest. Cover in severe cold.	
Cereals	Plantain, ribwort plantain, lucerne, sainfoin, esparsett, purple deadnettle, cornflower, a little amount of chamomile. Also, see the different types of cereals.				
Chamomile	Cabbage, onion. Promotes the growth of cabbage.	Mint, squash			
Chard	Bush bean, cabbage, onion, carrot, parsnip, radish, black radish, garlic.	Climbing bean, spinach.	Relatively easy to grow. Grows best in a deep, humus-rich and well-fertilized soil. So directly on the open land, 1-2 cm deep when the earth has become a little heated. Early sowing in cold soil increases the number of log runs.	Break off the outer leaves at the base. The thick stalks can be cooked like asparagus. Chard can be harvested well into the fall. Parboil before freezing. Small fresh leaves are good raw.	
			10-20 cm between the plants and 35-50 cm between the rows.		
Chervil	Lettuce. Attracts snails.				
Chilli	Basil, onions, mint, rosemary, scallion.	Dill			
Chinese cabbage		Corn			
Chive	Dill, fruit trees, early carrot, roses.	Beans, peas.			

Plant	Thrives with	Thrives Not with	Soil, Environment and Sowing	Harvesting	Problems and Solutions
Climbing bean	Cauliflower, cucumber, coriander, rutabaga, lettuce, sunflower, scorzonera, tomato.	Asian leafy vegetable, chive, fennel, onion, chard, leek, beetroot, scallion, white cabbage, garlic, pea.			
Clover	Primarily white clover provides organic nitrogen and helps keep the weeds away.		Soil improvers, green manure.		
Common Basil	Chilli, cucumber, oregano, asparagus, tomato, cabbage. Protects tomato and cucumber in the greenhouse against white flyers and aphids.	Wormwood, chard, sage			
Common rue		Lettuce, squash, white cabbage.			
Coriander	Anise, beans, spinach, peas. Anise and coriander together make them both germinate faster.	Fennel, squash			
Corn	Bean, potato, different kinds of cucumbers, sunflower, dill, pea. Plant beans around, however, the corn must have a few week's leads.	Celery Chinese cabbage, tomato.	Prefer deep sand blended earth. Needs a warm place. Never sow the corn in single or double rows as it will cause poor pollination. It is best to sow on a square surface so that many plants come next to each other. Sow about 4 cm deep and grow indoors about 25 C. Lower the temperature after the seed has germinated. Planted out when the risk of frost is over. In warmer areas, it can be directly sown, but the soil must be at least 15 C in order for the seed not to rot.	Preferably remove the lateral shoots at the bottom of the plant for better growth power and harvest. Harvested when the silky threads wither completely, and the pistons are thick and blunt. A print with a nail on a corn grain should give a dense and milky white juice.	

Plant	Thrives with	Thrives Not with	Soil, Environment and Sowing	Harvesting	Problems and Solutions
Corn salad	Belongs to the valerian family.s		Sow approx—1 cm deep. On open land until late summer for a prolonged autumn harvest and wintering. Keep the soil moist during the germination (10-14 days).	Pinch off the whole rosette at the surface of the soil as soon as it becomes large enough to eat. Harvest far into late autumn. Preferably used fresh. Very durable. Vitamin C.	
Cranesbill	Rose				
Cress	Parsnip, radish, black radish. Loved by the aphids which is both an advantage and a disadvantage.				
Crown imperial	Rose				
Cucumber	Basil, beans, dill, fennel, Jerusalem artichoke, chress, cabbage, onion, corn, parsley, beetroot, lettuce, celery, climbing beans, garlic, peas. Dill attracts beneficial insects that eat cabbage worms, small beetles, and aphids. In greenhouses, basil protects against white flyers and aphids.	Broadbean, spices, potato, radish, black radish, tomato.	Thrives in warm, shallow, moist, and nutritious soil. The need for potassium is excellent. While too much nitrogen yields overwhelming foliage, poor fruiting, and fruits of lower quality. Compost and well manure are the best fertilizers.	Harvest before the frost.	
Curly Mint	Increases taste and vigour in cabbage plants.	Squash			
Dandelion	Different kinds of fruit trees. Preferably apple and plum.				

Plant	Thrives with	Thrives Not with	Soil, Environment and Sowing	Harvesting	Problems and Solutions
Different kinds of cabbage	Beet, leaf celery, broad-bean, bush bean, bean, dill, endive, cucumber, haricot, chamomile, curly mint, onion, wormwood, chard, mint, peppermint, carrot, common marigold, rosemary, Celeriac, beetroot, lettuce, sage, scallion, celery, sunflower, spinach, scorzonera, thyme, pea. Celery keeps cabbage flies away with its strong scent. Marigolds against cabbage butterfly larvae. Haricot verts protect against lice, cabbage fly and cabbage worm. The growth is promoted by chamomile, mint, sage, wormwood, rosemary, and bush beans. Dill attracts beneficial insects that eat cabbage worms, small beetles, and aphids. Leaf celery protects against cabbage butterfly. Curly mint and peppermint enhance the taste and vigour. White clover can reduce the attack of the cabbage flies but can also reduce the harvest.	Climbing bean, strawberry, cabbage, potato, leek, garlic.	Nutritious and moist soil. Dehydration weakens the plant and benefits insects and diseases. Keeping the soil constantly moist reduces the risk of attack.		Sprinkle wooden ash or stone flour on and around the plants repeatedly against earth flies and cabbage caterpillar. Co-grow to reduce infestation. Burn sick plants, do not put them in the compost.

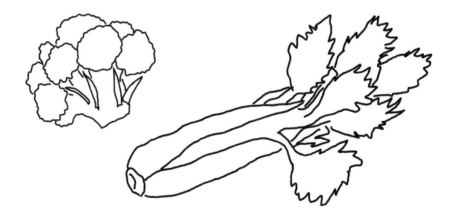

Plant	Thrives with	Thrives Not with	Soil, Environment and Sowing	Harvesting	Problems and Solutions
Different kind of fruit trees	Cyclamen, chive, nasturtium, cress, onions, dandelion, New Zealand spinach, mint, horseradish, rhubarb, tansy, marigold, thyme, white clover, garlic, white mustard, southernwood. Garlic against bud-eating birds and monilia. Chives and clover between the trees. Tansy, white mustard, and cyclamen help against parasites. Hang onion rings in the trees as the fruit ripens to deter the birds from eating the fruit. Dandelions are especially useful for apple and plum. Silverberry in fruit orchards can yield about 10% higher harvest. Particularly susceptible are nut-bearing trees and some plum trees.	Potato			
Dill	Broadbean, broccoli, bush bean, chive, cucumber, cabbage, onion, mint, parsley, beetroot, lettuce, scallion, asparagus, peas. Dill also attracts useful insects such as predatory spiders, golden eyed dragonflies, and predatory stink, which eats cabbage worms, small beetles and aphids among cabbage, cucumber, and lettuce. Perfect for pollinators.	Chili, fennel, carrot, tomato. Can be cross-pollinated with the carrot.	Slightly clayey soil for good access to phosphorus. Cultivated in sun/semi-shade. It should be 4-6 years between cultivation opportunities in the same place. Other herbaceous flowers - carrot, parsley, parsnip, etc. - must also not be grown there in the meantime.	At early harvest, the dill grows up again and gives a second harvest. Dill flowers are harvested when flowering is almost over.	When the plant turns brown and withers, it may be due to root fire or other propagating fungi. Prevents and combats with horsetail spade and algae extract.

Plant	Thrives with	Thrives Not with	Soil, Environment and Sowing	Harvesting	Problems and Solutions
Edible amaranth			Like hot weather, do not tolerate frost. Early sow indoors, planted out when the frost risk is over, and the soil has been warmer. It must not dry out and appreciate protection from cold weather. Once the plant has started, it grows quickly.	Young leaves and shoots are used, cooked, quick-fried, stewed, some- times raw, usually together with other vegetables. No dominant taste. A little fleshy leaf that is very nutritious. Especially on iron and calcium.	
Eggplant	Also, see the tomato. Protect potatoes against the Colorado beetle.	See tomato.			
Endive	Fennel, cabbage, onion, climbing bean. Conven- ient to sow at the place where you plan to grow turnip next year.				
Fennel	Cucumber, sage, peas.	Bush beans, dill, coriander, lettuce, climbing beans, tomato.			
Feverfew	Rose	Squash			
Flax	Potato				
Garden angelica	The angelica in the veg- etable garden attracts parasitic stings, ladybugs, stink bugs and other small, good friends.				
Garden orache	Radish.	Potato			
Garlic	Chili, fruit trees, cucum- ber, raspberry, strawber- ry, chard, carrot, rose, beetroot, tomato. Garlic is incredibly useful; its scent is said to scare away flying animals. Planted near plants that you want to protect is said to keep the vole away. Garlic water protects against aphids.	Broadbean, bean, cabbage, pea.			
Gooseberry		Potato	Prefer moist and well- drained soil. Can grow in light shade to direct sun.		
Goosefoot	See spinach	See spinach			

Plant	Thrives with	Thrives Not with	Soil, Environment and Sowing	Harvesting	Problems and Solutions
Haricot	Leaf celery, cucumber, strawberry, corn, potato.	Onion, garlic.	Soil improvers, green manure.		
Horseradish	Fruit trees. Suitable to plant around the potato garden.				
Hyssop	Rose, white cabbage. Good for bees	Squash			
Jerusalem arti-choke	Cucumber, Celeriac. Pro-tects potatoes, carrots, and other root vegetables against field vole. Plant as a shield outside the vegetable garden because it does not thrive with the potatoes.	Potato's			
Kohlrabi	Beet				
Lavender	Carrot, rose. It discour-ages flies and other bugs but attracts essential pollinators. One tip is a lavender hedge around the vegetable garden.	Parsley root, squash			
Leaf Celery	Snap beans, cabbage plants, leeks. Protects cabbage plants against cabbage butterfly		Early sow indoors; see Celeriac	Harvested before it gets too cold. Frost sensitive. With capping/coverage, the harvest season can be extended. Short shelf life, a few weeks, in a cold and humid place.	Spray cold coffee/ tea against aphids, can also be poured on the soil next to the plants.
Leek	Edible amaranth, leaf celery, broccoli, chilli, fruit trees, cucumber, strawberry, onion, carrot, lettuce, celery, scorzon-era, summer mallow (white), common mari-gold, tomato. The combi-nation of leaf amaranth, leek and summer mallow works well. Marigold is good against the leek moth.	Broadbean, bush bean, cabbage, chard, beetroot, climbing bean, pea.			

Plant	Thrives with	Thrives Not with	Soil, Environment and Sowing	Harvesting	Problems and Solutions
Lemon balm	Rose	Squash			
Lettuce	Bush bean, dill, cucumber, strawberry, cabbage, onion, chervil, carrot, peppermint, potato, leek, rhubarb, Celeriac, beetroot, scallion, climbing bean, scorzonera, thyme, pea. Dill attracts beneficial insects that eat cabbage worms, small beetles, and aphids. Chervil gives dense head and keeps the aphids away.	Fennel, common rue, lettuce, celery.			
Lilies	Rose				
Lupine			Soil improvers, green manure.		
Mallow	Rose				
Marigold	Cabbage, tomato. May help keep the cabbage butterfly larvae away from the cabbage.				
Melon	Potato				
Mint	Chilli, broccoli, cabbage, carrot, lettuce, tomato, pea. Promotes the growth of cabbage plants. The scent can confuse insects and keep the cabbage butterfly away from the cabbage.	Chamomile, parsley, squash			
Nasturtium	Fruit trees, potato, white cabbage.				
New Zealand spinach	Fruit trees				
Oats		Oats, the year before potatoes, can give potato scab.			

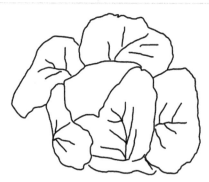

Plant	Thrives with	Thrives Not with	Soil, Environment and Sowing	Harvesting	Problems and Solutions
Onion	Asian leafy vegetables, cauliflower, brussel sprouts, chilli, dill, endive, fruit trees, cucumber, strawberry, cabbage, chard, carrot, leek, rose, Celeriac, radish, black radish, beetroot, lettuce, chamomile, tomato, scorzonera. Beetroot improves onion growth.	Broadbean, bean, potato, pea.	Sunny, warm, and airy location with soil rich of humus and sand. It can dry quite well.		
Oregano	Basil, white cabbage	Squash			
Paprika	See chilli	See chilli			
Parsley	Chilli, dill, cucumber, potato, radish, black radish, asparagus, tomato, pea. Roses get more scent if parsley is planted around the root.	Lavender, mint, lettuce.	Can be grown in most soils. Sow early about 1 cm deep. Grows very slowly and can advantageously put the seeds in lukewarm water a day before sowing. Wipe them off and then sow. Watering in the wound also helps.	The flat-leafed is tastier. Parsley can winter. Tastefully fresh but can be frozen or dried with an ok result, especially the flat-leafed.	
Parsley root	Jerusalem artichoke, carrot, potato, asparagus, tomato.	Lavender	Also, see parsnip, but the parsley root would rather have sand-blended and humus-rich soil. Sow 1 cm deep, early spring or late autumn.	Very cold resistant. If they are wintering, they must be picked as soon as the teal has gone out of the soil; otherwise, they will become woody and untasty. The leaves are used as regular parsley. Harvesting and storage as a parsnip.	

Plant	Thrives with	Thrives Not with	Soil, Environment and Sowing	Harvesting	Problems and Solutions
Parsnip	Borage, cress, onion, chard, radish, sweet corn.	Potato, carrot. Carrot can spread the carrot fly.	Deep, humus-rich and slightly more nutritious soil. Low nutrient supply gives hairy roots, and dry soil gives cracked roots. Sow 1-2 cm deep. To speed up otherwise slow germination (3 v), the seeds can be soaked for 24 hours in lukewarm water, dried and then sown. Always use fresh seeds and sow early. Can also be sown late autumn. The soil should not dry out during the germination.	Harvest as late as possible. Leave 2 cm of the tops and store as carrots. Frost resistant. Gets much sweeter after a proper frost. Overwintered parsnip contains three times more sugars than those harvested before winter. If they are wintering, they must be picked as soon as the teal has gone out of the soil; otherwise, they will become woody and distasteful.	
Pea	Broadbean, dill, fennel, cucumber, strawberry, coriander, cabbage, corn, carrot, mint, parsley, radish, black radish, lettuce. Carrots stimulate growth.	Bean, chive, onion, potato, leek, scallion, climbing bean, tomato, garlic. Peas the year before potatoes can give potato scab.	Soil improvers, green manure.		
Peppermint	Fruit trees, cabbage, carrot, rose, lettuce, tomato. Increases taste and vigour in cabbage plants. Its scent disturbs many insect pests.	Potato, squash			

Plant	Thrives with	Thrives Not with	Soil, Environment and Sowing	Harvesting	Problems and Solutions
Potato	Asian leafy vegetables, cauliflower, broadbean, haricot, bush bean, dill, nasturtium, coriander, cumin, turnip, corn horseradish, parsley, tansy, common marigold, Celeriac, lettuce, celery, spinach, marigold, eggplant. Eggplant against the Colorado beetle, farm beans against aphids, dandelion gives disease resistance. Soybeans the year before giving less scab. Coriander and cumin enhance the taste of the potato. Horseradish favours growth, but plant it either around the garden or in perforated pots that pass-through moisture but prevent the horseradish from spreading.	Birch, fruit trees, cucumber, raspberry, Jerusalem artichoke, gooseberry, cabbage, onion, melon, carrot, peppermint, pumpkin, rosemary, celery, sunflower, squash, tomato, garden orache, wheat, pea. Sunflower slows down growth. Peas, oats, and barley should not be grown the year before as they increase the risk of scab. Good with strawberries the following year, but not beans, peas, or beets.			
Pumpkin	Bean, dill, borage, chress, pea. Also, see the cucumber. Especially the cress and cucumber are good. Dill attracts beneficial insects that eat cabbage worms, small beetles, and aphids.	Broadbean, spicy herbs, potato, tomato	Thrives in warm, airy, moist, and nutritious soil. The need for potassium is great. While too much nitrogen yields overwhelming foliage, poor fruiting, and fruits of inferior quality. Compost and well manure are the best fertilizers.	Harvesting before the frost damages the fruits.	
Radish	Bush bean, strawberry, chress, cabbage, onion, chard, carrot, parsnip, parsley, beetroot, tomato, pea.	Cucumber, hyssop, spinach, squash, tomato.			
Raspberry	Spinach, garlic.	Potato	See the currant		
Red onion	Beet, fruit trees.				

Plant	Thrives with	Thrives Not with	Soil, Environment and Sowing	Harvesting	Problems and Solutions
Rhubarb	Bush bean, fruit trees, cabbage (early) lettuce, spinach. However, rhubarb thrives best alone.	Lucerne	Thrives best in a colder climate. Well-fertilized, moisturizing, and well-drained soil. Goodly cover the soil with cut leaves. The leaves also protect against infestation if you put them around other plants. Rot fungi can destroy the root in a water-sick and acidic soil.	Since rhubarb leaves full harvest only after three years, it cannot be included in the garden plant succession plan	
Rose	Boxwood, lemon balm, chive, borage, hyssop, crown imperial, lavender, lilies, onion, mallow, feverfew, cranesbill, geranium, peppermint, tansy, ajuga reptans, yarrow, sage, spinach, scarlet fuchsia, alyssum, lobularie, common marigold, thyme, scarlet beebalm, valerian, garlic. Onions provide stronger colour and more intense scent. Garlic increases the aroma of the rose, oil content and fragrance. Peppermint, lavender, sage, thyme, and hyssop protect against aphids. Chives give better scent, less sensitivity to black spot disease. Parsley makes the rose smell more if it is planted next to the root of the rose. Dig down some rusty nails next to the root to avoid insect infestation and mildew. Geraniums attract creeps that would otherwise damage the roses.				
Rosemary	Bean, chilli, cabbage, carrot, sage. Promotes the growth of cabbage plants and keeps insects away.				

Plant	Thrives with	Thrives Not with	Soil, Environment and Sowing	Harvesting	Problems and Solutions
Runner bean	Beet, strawberry, rutabaga, celery, scorzonera	Chive	Best in soil fertilized generously with compost. The soil should be warm during sowing, but then the beans will do well even if the summer is cool. Must be pollinated by insects.	Do not allow the beans to become too large.	
Rutabaga	Beet, runner bean, dill onion, chard, lettuce, spinach, climbing bean, tomato, pea.				
Rye	See also cereals. Cornflower and rye brome increase harvest.				
Sage	Fennel, cabbage, carrot, rose, rosemary. Promotes the growth of cabbage plants. Attracting pollinators.	Cucumber, scallion, squash			
Scarlet beebalm	Rose. Good for bees	Squash			
Scorzonera	Bean, cabbage (also kohlrabi) onion, carrot, leek, lettuce, spinach. Beneficial effect on onion, spinach, and lettuce. Is very deep-rooted can advantageously be co-grown with shallow-rooted plants. In the crop rotation, it is placed near carrots, which have about the same nutritional requirements.	Due to the danger of root eel, the scorzonera should not be grown in places where tomatoes or carrots have previously grown.			
Silverberry	Silver bushes in fruit orchards can yield about 10% higher harvest. Particularly susceptible are nut-bearing trees and some plum trees.				
Southernwood	Fruit trees.	Squash			
Soybean	The soybean thrives with all types of vegetables. Soybeans the year before potatoes give less potato scab.				

Plant	Thrives with	Thrives Not with	Soil, Environment and Sowing	Harvesting	Problems and Solutions
Spinach	Broadbean, bush bean, raspberry, strawberry, cabbage, rhubarb, rose, scorzonera.	Chard, beetroot.			
Squash	See pumpkin	See pumpkin.			
Stinging nettle	Tomato				
Stinking Roger	Cauliflower, fruit trees, potato, rose, tomato, white cabbage. Helps plants against nematodes by producing a nematode poison. Help keep white flyers in the greenhouse, especially on the tomatoes.				
Strawberries	Beans, borage, onion, leek, radish, black radish, lettuce, spinach, garlic, peas. Garlic protects against mites.	Asian leafy vegetables, cabbage, tomato.			
Subterranean clover	Cabbage. Protects against lice, cabbage fly and cabbage worm.		Soil improvers, green manure.		
Sunflower	Cabbage, corn, climbing bean.	Potato			
Tansy	Bramble, fruit trees, potato, rose, white cabbage. Protects fruit trees against parasites.				
Tarragon	White cabbage	Squash			
Thistle	Indicator Plant. The thistle shows that the soil structure is very poor, for example, for hard compression. The thistle's extraordinarily strong roots loosen the soil and improve it for next year's plants. It is crucial to remove the thistle flowers when they are at their best to avoid seed dispersal.				
Thyme	Fruit trees, cabbage, carrot, rose, lettuce, white cabbage. Keeps the cabbage worm away. Thyme attracting pollinators and mites.	Squash			

Plant	Thrives with	Thrives Not with	Soil, Environment and Sowing	Harvesting	Problems and Solutions
Tomato	Basil, bush bean, bean, borage, onion, corn, carrot, nettle, peppermint, parsley, leek, common marigold, parsley root, celeriac, radish, black radish, scallion, sage, asparagus, marigold, garlic. Sage for better growth and taste. Borage for earlier harvest. Red-leaved basil also stimulates the taste. Common basil protects tomatoes in greenhouses against white flyers and aphids. Nettles for more extended durability and against mould. Mexican marigold against white fliers, common tagetes protects against nematodes. Marigold for a better harvest. Fish cleaning from lake fish in the planting hole. Surround the tomatoes with beans to promote the growth of both.	Beet, dill, fennel, cucumber, strawberry, kohlrabi, cabbage, woodworm, potato, squash, black walnut, pea.			
Tuber fennel	Best grown separate.	Most kitchen and spice plants do not like fennel.	Thrives best in sheltered and warm locations on light soils. Sow 1-2 cm deep 5-6 weeks indoor before planting or in warm position when the soil holds at least 10 C. Handle the plant carefully when planting as damage to the roots increases the risk of flowering.	Harvested step by step. Cupping prolongs the harvest. The tubers can be stored in the fridge for a few weeks. Otherwise, it is stored through parboiling and freezing. The leaves are used as a spice.	Attacks are unusual, but they can be attacked by the same insects that attack carrot and parsley

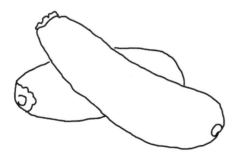

Plant	Thrives with	Thrives Not with	Soil, Environment and Sowing	Harvesting	Problems and Solutions
Turnip	Bean, pea, potato, carrot, beetroot, celery, fast-growing lettuce.		Sow early as soon as the soil is frost-free. Sow every fourteenth day of constant harvest all summer. Avoid excessive shadow. It thrives best in light, relatively nutritious and humus soil. A lot of organic matter in the soil gives rapid root growth, which is desirable to get excellent and juicy roots. After 7-8 weeks, you can start harvesting. Therefore, turnips are suitable as a between crop where there is a temporary vacancy in the garden.	You harvest turnips young when they are mild in taste. In any case, they should not be bigger than a tennis ball.	Like all other cruciferous plants, turnips are also attacked by earth fleas. They can occur in large quantities in dry and warm weather. Therefore, make sure the soil is always moist. You can also use ash or plain flour sieved or powdered over the leaves to prevent flea infestation.
Valerian	Rose	Squash	Develops an enormously, powerful root system.		
Walnut	Mulberries can work; otherwise, the walnut should grow itself as the tree lives in symbiosis with fungi that poison the soil.				
Wheat	Broadbean, pea, spinach, clover, soybean, chamomile, cockleboat, white mustard. Also, see cereals.	Potato, rye	For 100 kg seed - mix in 1 g of chamomile, 1 g of cockleboat and 1 g of white mustard.		
White clover	Fruit trees, cabbage. Can protect cabbage plants from cabbage flies, generally, as ground cover and for green manure.		Soil improvers, green manure.		
White mustard	Helps fruit trees against parasites.	White cabbage.			
Wild thyme	Good for bees				
Wormwood	Cabbage, white cabbage, carrot. Promotes the growth of cabbage plants.	Squash			

Plant	Thrives with	Thrives Not with	Soil, Environment and Sowing	Harvesting	Problems and Solutions
Yarrow	Indicator plant. Where the yarrow grows, there is a lack of potassium. The yarrow generates potassium which is returned to the earth as the yarrow decomposes				

Plants that indicate soil types

Lime-rich soil	Low pH	High pH	Good nutritious soil	Hard packed or poorly drained	Light, sandy soil
Wild Mustard - Sinapis arvensis	Corn Marigold - Glebionis segetum	Creeping Bellflower - Campanula rapunculoides	Nettle - Urtica dioica	Coltsfoot - Tussilago farfara	Common Storksbill - Erodium cicutarium
Speedwell - Lucus Veronicae	Corn Spurry - Spergula arvensis	Crow Garlic - Allium vineale	Ground Elder - Aegopodium podagraria	Long-Head Poppy - Papaver dubium	Corn Spurry - Spergula arvensis
Cabbage Thistle - Cirsium oleraceum	Pale Knotweed - Persicaria lapathifolia		Cow Parsley - Anthriscus sylvestris	Field Horsetail - Equisetum arvense	Hare's-foot Clover - Trifolium arvense
Wild Thyme - Thymus serpyllum - Wild Thyme	Meadow Buttercup - Ranunculus acris	Common Poppy - Papaver rhoeas - Common Poppy	Common Mugwort - Artemisia vulgaris	Creeping Buttercup - Ranunculus repens	Small Bugloss - Anchusa arvensis
Bloody Geranium - Geranium sanguineum - Bloody Geranium	Common Rush - Juncus effusus	Common Agrimony - Agrimonia eupatoria	Dandelion - Taraxacum	Broadleaved Plantain - Plantago major	Loose Silkybent - Apera spica-venti
Common Poppy - Papaver rhoeas - Common Poppy	Common Ragwort - Jacobaea Vulgaris	Erect Brome - Bromus Erectus	Common Groundsel - Senecio vulgaris	Curly Dock - Rumex Crispus	Rosebay Willowherb - Chamaenerion angustifolium
Cikoria - Cichorium intybus	Corn Poppy - Papaver dubium - Corn Poppy		Hoary Cress - Lepidium draba	Thistle - Cirsium	Sun Spurge - Euphorbia helioscopia
Field Bindweed - Convolvulus arvensis - Field Bindweed	Rockcress - Arabidopsis arenosa		Wormseed Mustad - Erysimum cheiranthoides		Thrift - Armeria maritima
	Chamomile - Chamomilla recutita		Field Pennycress - Thlaspi arvense		True Forget-me-not - Myosotis scorpioides
	Field Pennycress - Thlaspi arvense		Fat hen - Chenopodium album		Corn Marigold - Glebionis segetum
	Lady's Bedstraw - Galium verum		Perennial Sowthistle - Sonchus arvensis		Common Fumitory - Fumaria officinalis
			Creeping Yellowcress - Rorippa sylvestris		

Nitrogenous soil	Potassium-rich soil	Potassium-poor soil	Soil low in humus	Soil high in humus	Moist or water-sick soil
Nettle - Urtica dioica	Fat hen - Chenopodium album	Black Medick - Medicago lupulina	Cornflower - Centaurea cyanus	Annual Nettle - Urtica urens	Silverweed cinquefoil - Argentina anserina
Large-Flowered Hemp Nettle - Galeopsis speciosa	Buttonweed - Malva neglecta	Brown Knapweed - Centaurea jacea	Corn Marigold - Glebionis segetum	Large-Flowered Hemp Nettle - Galeopsis speciosa	Water-pepper - Persicaria hydropiper
Black Nightshade - Solanum nigrum	Strawberry Clover - Trifolium fragiferum	Red Clover - Trifolium pratense	Common Nipplewort - Lapsana communis	Fat hen - Chenopodium album	Meadow Buttercup - Ranunculus acris
Curly Dock - Rumex Crispus		Common Yarrow - Achillea millefolium	Sun Spurge - Euphorbia helioscopia	Wild Garlic - Allium ursinum	Wild Mint - Mentha arvensis
Good-King-Henry - Blitum bonus-henricus			Corn Gromwell - Lithospermum arvense	Curly top Knotweed - Persicaria lapathifolia	Tufted Hairgrass - Deschampsia cespitosa
White Nettle - Lamium album			Common Knotgrass - Polygonum aviculare		Common Chickweed - Stellaria media
Herb Robert - Geranium robertianum					Pilewort - Ficaria Verna
Common Chickweed - Stellaria media					Scentless False Mayweed - Tripleurospermum inodorum

pH around 4,4	pH around 4,7	pH around 5,0	pH around 5,3	pH around 5,6	pH around 6,2
Lapland Cornel - Cornus suecica	Tormentil - Potentilla erecta	Brittle Bladder-fern - Cystopteris fragilis	Allium 'Mercurius' - canis est Mercurius	Baneberry - Actaea spicata	Wood Sanicle - Sanicula Europaea
Chickweed-Wintergreen - Trientalis Europaea	Fern - Pteridophyta	One-flowered Wintergreen - Moneses uniflora	Scotch False Asphodel - Tofieldia pusilla	Cowslip - Primula veris	Common Twayblade - Neottia ovata
Common Oak Fern - Gymnocarpium dryopteris	Yellow Archangel - Lamium galeobdolon	Sweet-scented Bedstraw - Galium odoratum	Water Avens - Geum rivale	Yellow Wood Anemone - Anemone ranunculoides	
False Lily - Maianthemum bifolium	Wood Stitchwort - Stellaria nemorum	Wood Cranesbill - Geranium sylvaticum	Butterfly Orchid - Platanthera bifolia	Lesser Celandine - Ficaria verna	
European Goldenrod - Solidago virgaurea	Bearded Wheatgrass - Elymus caninus	Club Spikemoss - Selaginella selaginoides	Woodland Angelica - Angelica sylvestris		
Wood Sorrel - Oxalis acetosella	Long Beech Fern - Phegopteris connectilis	American Milletgrass - Milium effusum	Common Lady's Mantle - Alchemilla Vulgaris		
Raspberry - Rubus idaeus	Wood Anemone - Anemone nemorosa	Marsh Horsetail - Equisetum palustre	Red Campion - Silene dioica		
Twinflower - Linnaea borealis	Alpine Sow-thistle - Cicerbita Alpina	Coral Root - Cardamine bulbifera	Fumewort - Corydalis		
Common Cow-wheat - Melampyrum pratense	Lily-of-the-valley - Convallaria majalis	Yellow Star Of Bethlehem - Gagea lutea	Nettle - Urtica dioica		
		Wood-sedge - Carex digitata	Hedge Woundwort - Stachys sylvatica		
		Spring Vetchling - Lathyrus vernus	Marsh Grass - Parnassia palustris		
		Marsh Thistle - Cirsium palustre	Wild Garlic - Allium ursinum		
		Wall Lettuce - Lactuca muralis	Ground Elder - Aegopodium podagraria		
		Wild Strawberry - Fragaria vesca	Cudweed - Galium triflorum		
		Rough Horsetail - Equisetum hyemale	Meadowsweet - Filipendula ulmaria		
		Common Globeflower - Trollius europaeus	Jerusalem-sage - Pulmonaria		
		Common Sorrel - Rumex acetosa	Herb Paris - Paris quadrifolia		
		Hawksbread - Crepis paludosa	Common Hepatica - Anemone hepatica		
		Melancholy Thistle - Cirsium heterophyllum	Mezereum - Daphne mezereum		

Optimally pH 4-5-6,5 Can withstand higher pH less well	pH insensitive genus	Optimally pH 4,5 - 6,5 Withstands higher PH very well
Maple - Acer	Horse-chestnut - Aesculus hippocastanum	Ash - Fraxinus excelsior
Birch - Betula	Alder - Alnus	Sweet cherry - Prunus avium
Oak - Quercus robur	Beech - Fagus sylvatica	Pussy willow - Salix caprea
Shadbush - Amelanchier spicata	Rowan - Sorbus aucuparia	Willow - Salix
Beautybush - Kolkwitzia Amabilis	Cornus - Cornus	
Common Privet - Ligustrum vulgare	Holly - Ilex aquifolium	
Fly Honeysuckle - Lonicera xylosteum	Snowberry - Symphoricarpos albus	

Plants that thrive in shade

Leaf vegetables and cabbage plants
Many leaf plants feel good of a little shade because then they grow slower, which means longer harvest time. Instead, full sun can cause them to grow too quickly, become lankier and bloom faster.

Name	Use	Note
Leaf celery	To get paler stems.	
Cauliflower	Can handle half shade if it is well-drained. Otherwise in the sun	
Cabbage	Most cabbage plants become brittle and delicate to "leaf green".	
Chard	Both more beautiful and fuller.	
Lettuce	All varieties prefer shadier locations.	
Chinese mustard	The leaves become softer and milder in shadow positions.	
Pac Choi	Works well in the shade.	Applies to most other Asian leafy vegetables.
Rocket salad	Does not get as bitter in shadow	
Spinach	Only needs 4 hours of sun a day.	
Watercress	Can grow in the shade without water.	

Onion plants and herbs
Onion plants and herbs do not get as good in semi-shade, but you do not have full sun available so grow anyway! Several herbs can withstand half-shade, although they can be a bit lanky and not as flavorful, especially the herbs that are woody. Same with onion plants that in the shade do not develop their onions as well, but the tops grow well.

Name	Use	Note
Lemon balm	Is easily burned and dries out too quickly in full sun.	
Chive	Can last several years.	
Sand leek	It thrives best in a grove area and semi-shade.	
Coriander	Grows well and the plant easily self-seeds.	
Mint	Thrives best in moist soil. All varieties thrive better in semi-shade.	
Oregano	Works in semi-shade, but becomes lankier and softer in taste.	The yellow variant thrives better in semi-shade because it is quickly burned in bright sunlight.
Parsley	Grows well as a biennial plant.	
Ramson	It thrives best in a grove area and semi-shade.	
Scallion	If it is too shady, the onion does not develop so well but is grown instead for the tops.	
Garlic	If it is too shady, the onion does not develop so well but is grown instead for the tops.	

Root vegetables

Some root vegetables may develop in semi-shade. Others are grown in semi-shade like "baby leaf" for salads.

Name	Use	Note
Jerusalem artichoke	Can grow in half shade if it gets sun at least 4 hours/day.	
Carrot	Cultivated as baby carrots in the first place.	
Parsnip	See carrot.	
Potato	Pre-germinated, really early varieties work.	
Radish	Works well because it grows so fast.	
Beetroot	The root develops poorly. Small leaves for salad or use the tops as spinach.	

More plants for the shadow

Finally, some other plants that can handle some shade.

Name	Use	Note
Bean	Best developed when cooler. About 4 hours of sun/day is sufficient.	Not broadbean.
Ground elder	Grows everywhere.	
Nettle	Works well.	
Rhubarb	Works well in semi-shade.	
Wild strawberry	Works in wandering shadow.	
Asparagus	Is not as light-demanding because it is the shots that are eaten.	
Black currant	Does not require full sun all day.	
Pea	As the bean the peas like a little cooler climate.	

Plants for wind protection

	Wind strong 1-5	Storm solid 1-5	Early leaf cracking 1-5	Late leaf cracking 1-5	Pioneer, Secondary plant	Adult height in meters	Food plants (bee, bird, field and horn animals)	Exposure to wildlife attacks	Root and stump shot	1 Moist - 5 Dry	1 Low in nutrition - 5 Nutritious
TREES WITH LEAFS											
Acer campestre - Field Maple	4	3	2	3	P,S	8-10	-	3	st	3-5	4-5
Acer platanoides - Norway Maple	3	3	3	2	S,P	15-20	be	3	st	3-4	3-5
Aesculus hippocastanum - Horse-chestnut	2	2	4	2	S,P	15-20	-	-	-	3-5	4-5
Alnus glutinosa - European Alder	3	2	3	4	P,S	15-20	be, bi	1	st	1-3	3-5
Alnus incana - Gray Alder	3	2	2	3	P	8-12	be, bi	1	ro, st	2-4	3-5
Betula pendula - Warty Birch	2	2	3	2	P	20-25	bi	1	st	3-5	1-3
Carpinus betulus - Hornbeam	2	2	3	4	S	12-15	bi	4	st	3-5	3-5
Fagus sylvatica - Beech	2	2	3	5	S	20-25	bi	4	st	3	2-4
Fraxinus excelsior - Ash	3	4	1	2	P,S	20-25	-	4	st	1-3	4-5
Malus sylvestris - Crabapple	1	2	2	3	S,P	6-9	be, bi, fa, ha	5	st	3	3-4
Populus x canescens - Grey Poplar	4	4	-	-	P	15-20	-	-	ro, st	2-5	2-4
Populus tremula - Aspen	3	3	3	2	P	10-30	p, ha	-	ro, st	2-5	2-4
Prunus avium - Wild Cerry	1	2	3	4	S,P	15-20	be, bi	3	ro, st	3	3-5
Prunus padus - Bird Cherry	3	2	4	2	P,S	8-12	be, bi	3	st	1-2	2-4
Prunus serotina - Black Cherry	2	2	-	-	S,P	8-10	be, bi	-	st	3-4	2-3
Quercus petraea - Sessile Oak	4	3	1	4	S,P	20-25	-	-	st	3-5	2-3
Quercus robur - Common Oak	4	3	2	3	S,P	20-25	bi, ha	3	st	3-5	2-4
Salix alba - White Willow	4	3	4	2	P	15-20	be, ha	3	st	1-2	4-5
Sorbus aucuparia - Rowan	3	1	4	4	P,S	8-12	be, bi, ha	3	st	3-5	2-3
Sorbus intermedia - Swedish White-beam	4	4	3	4	S,P	12-15	be, bi	5	st	3-5	3-5
Tilia cordata - Small-leaf Lime	2	2	2	2	S	20-25	be	2	ro, st	3	3-5
SHRUB TREES WITH LEAFS											
Crataegus monogyna - Common Hawthorn	4	4	3	5	S,P	4-7	be, bi, fa	3	st	3-4	3-5
Elaeagnus Angustifolia - Russian Olive	4	-	2	3	P,S	3-6	-	-	-	3-5	3-5
Hippophae rhamnoides - Sea-buckthorn	4	3	3	4	P	3-6	bi	3	ro, st	3-5	3-5
Prunus cerasifera - Cherry Plum	2	2	-	-	P	7-9	-	3	-	3-5	3-5
Sambucus nigra - Elder	4	4	4	4	P,S	3-6	bi	1	ro, st	1-3	4-5

	Wind strong 1-5	Storm solid 1-5	Early leaf cracking 1-5	Late leaf cracking 1-5	Pioneer, Secondary plant	Adult height in meters	Food plants (bee, bird, field and horn animals	Exposure to wildlife attacks	Root and stump shot	1 Moist - 5 Dry	1 Low in nutrition - 5 Nutritious
SHRUBS WITH LEAFS											
Amelanchier sp - Shadbush	3	3	-	-	S,P	5-7	be, bi	-	st	3	3-4
Caragana arborescens - Siberian Pea Tree	3	1	1	2	P,S	3-5	bi	3	st	3-5	1-3
Cornus sanguinea - European dogwood	2	3	2	2	S,P	3-5	bi	2	st	1-4	4-5
Corylus avellana - European Filbert	1	2	3	4	S,P	5-8	-	3	st	2-3	3-5
Euonymus europaeus - European Spindle	1	1	4	4	P,S	3-5	be, bi	-	st	1-3	3-5
Ligustrum vulgare - Wild Privet	2	2	2	5	S,P	2-3	-	-	st	3-5	3-5
Lonicera xylosteum - Fly Honeysuckle	2	3	5	2	S,P	3-4	-	-	st	1-4	3-5
Malus sargentii - Sargents Crabapple	3	2	-	-	S,P	1-3	be, bi	5	st	2-3	3-5
Physocarpus opulifolius - Common Ninebark	1	2	4	4	P,S	2-4	-	-	-	2-3	2-4
Prunus spinosa - Blackthorn	4	3	2	2	P	2-4	be, bi	2	ro, st	2-3	3-4
Ribes alpinum - Alpine Currant	2	2	5	4	S,P	1-2	be, bi	1	st	1-3	4-5
Rosa canina - Dog Rose	3	3	3	2	P	2-4	be, bi	-	ro, st	3-4	3-4
Rosa rugosa - Rugosa Rose	5	5	3	3	P	1-3	be, bi	-	ro, st	3-5	2-3
Symphoricarpos albus - Common Snowberry	2	-	2	4	S	1-2	be, bi	-	ro, st	3-4	2-4
Syringa vulgaris - Common Lilac	3	3	4	-	S,P	3-5	-	3	ro, st	3-4	4-5
Viburnum opulus - Guelder Rose	2	2	2	4	S,P	2-4	be, bi	3	st	1-3	3-5
CONIFER											
Larix decidua - Common Larch	3	1	4		P	20-25	bi	3	-	2-4	2-4
Larix kaempferi - Japanese Larch	3	1	4	-	P	15-20	bi	3	-	2-4	2-4
Picea glauca - White Spruce	5	5			S,P	12-15	bi	1	-	2-3	2-4
Picea sitchensis - Sitka Spruce	5	5			S,P	15-20	bi	1	-	1-3	2-4
Pinus mugo - Mugo Pine	4	4			P,S	6-8	bi	3	-	3-5	1-3
Pinus nigra - Black Pine	3	3			P	20-25	bi	3	-	2-5	3-5
Pinus sylvestris - Scots Pine	3	1			P	10-30	bi	3	-	3-5	1-3

The two plants to be extra careful with in connection with agricultural cultivation are different species in the barberry genus as well as bird cherry trees. Barberry is the host plant for a black roast that attacks wheat, and the bird cherry trees are the host plant for an aphid that attacks oats. For the black rust, the fungus only spreads within the immediate area while aphids can drift with winds a longer distance. Attack of hawthorn rust is another problem that can be avoided if all Chinese junipers in the immediate area are removed.

Alphabetical list of contents

ÄNGSVÄGEN - AN EXAMPLE

Alphabetical list of illustrations

My notes